Unveiling the Brain's Nanoscale Secrets: A Look Inside Synaptic Function

Zahara

Table of Contents

1. Introduction

The brain is composed of billions of neurons that form trillions of synaptic connections. The chemical synapse represents the basic anatomical and functional unit underlying information transfer in the brain. Complementary and multidisciplinary approaches are required to achieve a comprehensive understanding of how synapses work at the level of their structural architecture, their molecular organization, and how their behavior is dynamically regulated during the process of information transfer. In neuronal networks in the brain, plastic changes in synaptic transmission efficacy are hypothesized to underlie complex and precise processes such as sensory processing, motor control, memory, and cognition (Purves et al., 2004).

My doctoral work particularly focuses on the relationship between synaptic ultrastructure and function. Although many model synapses have been functionally and morphologically characterized, it remains unclear how the structural organization of a synapse at the level of individual active zone release sites contributes to its functional properties. Electron microscopy (EM) remains the gold standard for accurately resolving the subcellular organization of lipid-bound synaptic organelles (i.e. synaptic vesicles) and synaptic subcompartments (i.e. active zone release sites). Recent studies have indicated that synapse function is profoundly influenced by subtle differences in the structural organization of synapses operating in the nanoscale range (Imig et al., 2014; Siksou et al., 2009a). Additionally they have emphasized that the detection of these changes is critically dependent on tissue preparation protocols for ultrastructural analysis (i.e. cryo-fixation) and on the imaging techniques used to resolve synaptic ultrastructure (i.e. electron tomography) (Imig et al., 2014; Siksou et al., 2009a).

My book work takes into account these important methodological considerations in a comparative ultrastructural analysis of hippocampal Schaffer collateral and mossy fiber synapses. These experiments are designed to provide an unprecedented, high-resolution perspective of active zone organization at the mossy fiber synapse, one of the most functionally and morphologically enigmatic synapses in the brain. Corresponding analyses of Schaffer collateral synapses, which represent arguably the most extensively characterized synapse in the brain, are designed to provide a reliable morphological and functional frame of reference.

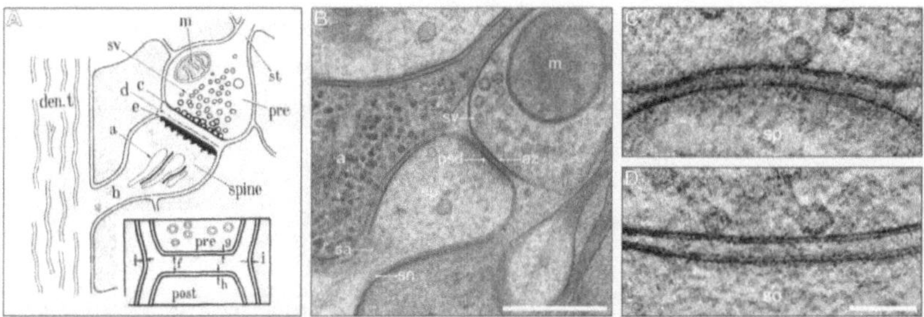

Figure 1. Synaptic ultrastructure.

(**A**) Schematic representing the first electron microscopic characterization of a small cortical spine synapse and its respective subcellular compartments. (**B**) Transmission electron micrograph of a small Schaffer collateral spine synapse in an organotypic hippocampal slice culture prepared by high-pressure freezing and freeze substitution (Imig and Cooper, 2017). (**C** and **D**) Characteristic features of asymmetrical (**C**; Grays Type I; excitatory) and symmetrical (**D**; Grays Type II; inhibitory) synapses. Docked vesicles are indicated with green arrowheads. Abbreviations: den.t, dendrite; sv, synaptic vesicle; sa, spine apparatus; sn, spine neck; sp, spine; so, soma; m, mitochondrion; az, active zone; a, astrocyte. Scale bars: 500 nm, **B**; 100 nm, **C** and **D**. Permission & Rights (**A**) from *Gray, 1959* with permission from Copyright Clearance Center (license number 4786470690034); (**B**) from *Imig and Cooper, 2017* with permission from Copyright Clearance Center (license number 4786471226892).

1.1. Synaptic ultrastructure and function: a historical perspective

Since its invention by Ernst Ruska in 1933 (Borries and Ruska, 1933; Ruska, 1933), the electron microscope has served as an invaluable research tool in the field of synaptic neurobiology. The structural composition of chemical synapses was first directly visualized in electron micrographs almost 65 years ago (Gray, 1959; Palay and Palade, 1955; De Robertis and Bennett, 1955). In his seminal study of synapses in the visual cortex of rats, Gray revealed a compartmental organization in which the axon of the signaling neuron terminated in a presynaptic bouton in close proximity to a postsynaptic compartment in the form of a dendritic spine (Figure 1 A and B) (Gray, 1959). Of critical importance to the understanding of synaptic function at the time, this ultrastructural view of the synapse revealed (i) that the axon terminal was filled with small vesicular organelles, termed synaptic vesicles, and (ii) that pre- and postsynaptic compartments were not in direct physical contact, but rather separated by what became known as the synaptic cleft. Subsequent studies performed in the frog neuromuscular junction built on this information by demonstrating that synaptic vesicles store and release chemical transmitter substances, fuse with the presynaptic membrane, and recycle to generate new vesicles during sustained activity (Ceccarelli et al., 1973; Heuser and

Reese, 1973). Together, these findings provided the basis for understanding one of the most fundamental and pervasive ultrastructure-function relationships in neurobiology, namely that synaptic vesicles are the morphological correlates of the quantal neurotransmission identified by Castillo and Katz in 1954 (Castillo and Katz, 1954).

Importantly, EM enabled the visualization of sites of trans-synaptic information transfer at a subcellular scale and linked with the functional properties of individual synapse subtypes. Electron dense "membrane thickenings" indicative of functionally specialized cellular subcompartments were observed at opposing pre- and postsynaptic membranes across the synaptic cleft (Figure 1 A and B) (Gray, 1959; Palay, 1956). The active zone, a term reflecting observations that synaptic vesicles preferentially cluster and fuse at this presynaptic specialization (Couteaux and Pécot-Dechavassine, 1970), was postulated to provide the molecular and structural environment required for the spatio-temporally regulated release of neurotransmitter into the synaptic cleft (Phillips et al., 2001; Triller and Korn, 1985). The postsynaptic density, located in direct apposition to the active zone, was analogously postulated to provide the molecular and structural environment required to cluster membrane-bound receptors capable of receiving a transmitter signal (Okabe, 2007). The observation that the ultrastructural appearance of synaptic active zones and postsynaptic densities correlated with respect to the anatomical location, neurotransmitter content, and behavioral properties of the synapse ultimately led to a classification system relating the morphological and functional characteristics of synaptic subtypes (Colonnier, 1968; Eccles, 1964; Gray, 1959; Uchinozo, 1965). Grays Type I, or asymmetric, later classified as excitatory, form synapses onto dendritic shafts or spines, harbor spherical presynaptic vesicles, and have a pronounced postsynaptic density (Figure 1 C); Grays Type II, or symmetric, later classified as inhibitory, innervate neuronal soma and dendritic shafts, harbor pleiomorphic presynaptic vesicles, and have comparably sized active zones and postsynaptic densities (Figure 1 D) (Colonnier, 1968; Eccles, 1964; Gray, 1959; Uchinozo, 1965). Further support for this classification system was subsequently provided by the demonstration that synaptic vesicles in Type I and Type II synapses were immunoreactive against the main excitatory neurotransmitter glutamate and the main inhibitory neurotransmitter γ-aminobutyric acid (GABA), respectively (Barbaresi et al., 2001; Beaulieu and Somogyi, 1990).

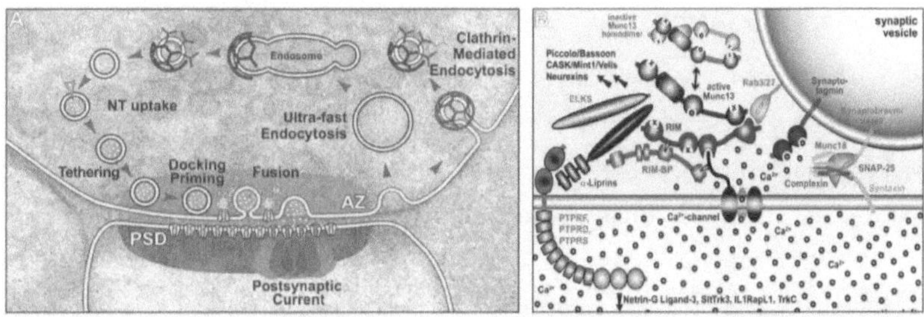

Figure 2. The synaptic vesicle cycle and molecular regulation of stimulus-evoked secretion at presynaptic active zones.

(A) Schematic of classical synaptic vesicle cycle. Newly synthesized neurotransmitters such as glutamate are actively transported from the cytoplasm into membrane-bound synaptic vesicles. Synaptic vesicles with transmitter cargo translocate to the plasma membrane where they dock and are molecularly primed for fusion with the plasma membrane upon calcium influx. Excess membrane caused by full collapse fusion with the plasma membrane is then recycled into the presynaptic terminal via both clathrin-dependent and clathrin-independent mechanisms. The recycled membrane is brought to endosomal structures where new synaptic vesicles are formed in a clathrin-dependent mechanism and then recycle back into the pool of synaptic vesicles. (B) Schematic of key regulatory molecules in excitation-secretion coupling at presynaptic active zones. These include: (i) active zone proteins such as ELKS, RIM-BP, and RIM that localize calcium channels to the presynaptic membrane and in proximity to synaptic vesicles, (ii) priming molecules such as Munc13s and Munc18 allow for synaptic vesicles to dock and prime at the plasma membrane bringing the vesicular and plasma membrane SNAREs close enough to interact, and (iii) SNARE complex components synaptobrevin, SNAP-25, and syntaxin, that catalyze the fusion of vesicular and plasma membranes. Abbreviations: NT, neurotransmitter; PSD, postsynaptic density; AZ, active zone. Permission & Rights (B) from Südhof, 2012 with permission from Copyright Clearance Center (license number 4862960297628).

The aforementioned studies demonstrated the major contribution EM has made towards our understanding of the fundamental principles underlying synaptic transmission, and emphasize that systematic morphological analyses provide functional insight, even on a single synapse level. Although excitatory and inhibitory synapses represent the majority of synapses in the central nervous system, a considerably broader spectrum of synapse classes, neurotransmitters, and receptor subtypes is ultimately required to support complex brain functions. Moreover, the observation that synapses of a given class (i.e. Grays Type I or II) or neurotransmitter subtype can differ substantially in their efficacy of evoked transmitter release (Purves et al., 2004) emphasizes the need for further investigation of ultrastructure-function relationships using refined methods and protocols designed to probe synaptic ultrastructure with higher stringency and to link electron microscopic observations with the molecular mechanisms underpinning fundamental synaptic properties.

1.2. Synaptic transmission

1.2.1. The synaptic vesicle cycle

Despite the broad spectrum of synaptic neurotransmitters, morphologies, and behaviors required to support complex brain functions, all chemical synapses operate by a stereotypic principle. The propagation of information from one neuron to the next requires both an electrical and a chemical component. An action potential occurs when a change in the electrochemical gradient across the semi-permeable plasma membrane of a neuron causes the opening of voltage-gated ion channels. This signal propagates rapidly along the axon of a neuron until it reaches a presynaptic terminal. Within the presynaptic terminal, voltage-gated calcium channels (VGCCs) open and the influx of calcium into the presynaptic terminal triggers the fusion of synaptic vesicles with the active zone membrane and the subsequent release of their lumenal cargo, chemical messengers, or neurotransmitters, into the synaptic cleft in a process termed exocytosis (Figure 2). Once the neurotransmitter binds to a postsynaptic ionotropic receptor, the cycle of signal propagation continues as evoked currents are integrated in the postsynaptic neuron.

Stimulus-coupled exocytosis is tightly regulated by a complex molecular machinery that operates at multiple steps preceding synaptic vesicle fusion to regulate the spatio-temporal precision of transmitter release (Figure 2 B). Experimental approaches combining mouse genetics with functional assays and corresponding ultrastructural analyses have identified key presynaptic protein components of this machinery and generated a view of the synaptic vesicle cycle (Figure 2 A) in which important regulatory steps can be related to the spatial organization of synaptic vesicles within the presynaptic terminal. This view illustrates a sequence of steps in which synaptic vesicles are: (i) filled with neurotransmitter, (ii) recruited to the active zone and loosely attached, or tethered, in proximity to the presynaptic membrane, (iii) docked in physical contact with the plasma membrane and rendered fusion-competent in a molecular priming step, (iv) and fused with the plasma membrane upon detection of elevated presynaptic calcium (Figure 2 A).

1.2.2. Synaptic vesicle tethering

The presynaptic compartment is ultrastructurally characterized by the accumulation synaptic vesicles at active zone release sites. Large, multivalent active zone scaffold molecules, including rab3-interacting molecule (RIM), bassoon, piccolo, and RIM binding protein (RBP) (Figure 2 B), create the local proteinaceous environment required to support stimulus-evoked neurotransmitter secretion. This protein network serves as a hub by mediating interactions between vesicular proteins (Geppert et al., 1997), VGCCs (Han et al., 2011; Südhof, 2012), and soluble components of the vesicle fusion apparatus (Figure 2) (Augustin et al., 1999; Brockmann et al., 2019; Südhof and Rizo, 2011). Although targeted genetic perturbations have implicated several components of the active zone scaffold in the recruitment of synaptic vesicles to active zone release sites (Han et al., 2011; Mukherjee et al., 2010; Südhof, 2012; Wang et al., 2016), a structural view of how this recruitment occurs is still evolving.

Experiments combining freeze-fracture with shallow etching provided early structural support for the concept that filamentous proteins play a role in the organization of synaptic vesicles at presynaptic active zones (Landis et al., 1988). Although rarely captured in aldehyde-fixed preparations, presynaptic filaments have since been described in a variety of species and synapse types prepared by rapid cryofixation, freeze substitution and plastic embedding (Bruckner et al., 2017; Cole et al., 2016; Siksou et al., 2009a; Stigloher et al., 2011; Vogl et al., 2015). More recently, cryo-electron tomographic reconstructions of frozen-hydrated synaptosomes have enabled detection and quantification of filaments both between vesicles ("connectors"), and between vesicles and the presynaptic plasma membrane ("tethers") (Fernández-Busnadiego et al., 2010, 2013). The observation that the number and length of tethers per vesicle is inversely related to active zone proximity supported a model in which vesicles are initially tethered by single, long filaments (>5 nm) before being anchored closer to the membrane by multiple short (<5 nm) tethers (Fernández-Busnadiego et al., 2010, 2013). Moreover, an analysis of RIM1α knock-out (KO) synapses revealed a perturbation of the organization of filaments and vesicles that implicated the active zone scaffold in these processes (Fernández-Busnadiego et al., 2013). However, the potential involvement of other active zone proteins in this phenotype must be considered, particularly in light of previous studies indicating that RIM is required for the correct active zone targeting of mammalian uncoordinated protein (Munc13) priming proteins (Andrews-Zwilling et al., 2006). The detection of filamentous connectors and tethers in other cellular subcompartments is also consistent with the notion that multiple protein species, or protein isoforms, contribute to the formation of vesicle-associated filaments (Hallermann and Silver, 2013; Schrod et al., 2018). Although their molecular identity remains to be determined, tethers provide a structural framework and level of organization that is likely critical for maintaining the supply of vesicles to the active zone during sustained activity (Hallermann and Silver, 2013).

1.2.3. Synaptic vesicle docking

In contrast to tethering processes, the molecular mechanisms responsible for docking synaptic vesicles in close contact with the presynaptic membrane are relatively well understood (Südhof and Rizo, 2011). To appreciate the functional significance of the synaptic vesicle docking process, an understanding of multiple overlapping concepts is necessary. Accordingly, in this section I will discuss the concept of a readily-releasable pool (RRP) of synaptic vesicles in context with the molecular machinery required to generate it and ultimately trigger its fusion with the active zone membrane during stimulus-evoked neurotransmitter release.

1.2.4. The readily-releasable pool of synaptic vesicles

In order to fuse in response to the arrival of an action potential, synaptic vesicles must undergo a molecular priming process that renders them fusion competent. It is this priming process that ensures a RRP of vesicles is available to fuse and release neurotransmitters in response to the arrival of an action potential. The RRP of a synapse is typically assessed by functional means and corresponds to the number of vesicles that fuse with the synapse in response to strong, vesicle-depleting stimuli (Kaeser and Regehr, 2017; Neher, 2015). A range of functional assays has been developed to accommodate the specific demands of different experimental preparations (Ariel and Ryan, 2010; Bekkers and Stevens, 1991; Neher and Marty, 1982; Rizzoli and Betz, 2005; Rosenmund and Stevens, 1996; Schikorski and Stevens, 2001; Schneggenburger et al., 1999). These and associated caveats will be discussed in more detail in the methodological considerations in the discussion (see section *4.3.1 Limitations of RRP estimates*). Major methodological differences include the type of stimulus applied to trigger transmitter release and the experimental means of detecting it. In low-density cultures, the rapid application of hypertonic sucrose reliably triggers fusion of the RRP (Bekkers and Stevens, 1991; Rosenmund and Stevens, 1996). However, this approach is not applicable to acute slice preparations where high frequency trains of action potentials are preferentially used to deplete the RRP (Schneggenburger et al., 1999; Thanawala and Regehr, 2013). Both approaches rely on a simultaneous measurement of postsynaptic responses to measure evoked transmitter release, which in glutamatergic neurons is manifest as excitatory postsynaptic currents (EPSCs). There are several caveats which limit the accuracy of RRP measurements obtained using these methods. These caveats include: (i) hypertonic sucrose-evoked RRP measurements are critically dependent on the speed at which the sucrose is delivered to the cell and the precise mechanism of operation remains poorly defined; (ii) the accuracy of RRP estimates calculated by back-extrapolation of cumulative EPSCs evoked during high frequency action potential trains is sensitive to both dynamic changes in release probability during the stimulus train (i.e. short-term plasticity) and RRP refilling (i.e. calcium-dependent priming); and (iii) postsynaptic responses report vesicle fusion indirectly and are consequently sensitive to dynamic changes in receptor properties (i.e. sensitization or saturation). Alternative approaches have been developed directly assay vesicle fusion. For example, the measurement of presynaptic capacitance evoked during the application of depolarizing voltage steps elicits dynamic changes in presynaptic membrane surface area resulting from evoked exo- and endocytosis (Delvendahl et al., 2016; Neher and Marty, 1982). However, this technique is only applicable to very large presynaptic boutons and is unable to discriminate concomitant exo- and endocytic processes during a depolarizing pulse. Alternatively, various optical approaches have been developed to report vesicle fusion, including lipid soluble fluorescent dyes (Rizzoli and Betz, 2004; Schikorski and Stevens, 2001), pH-sensitive fluorescent reporters fused to vesicular proteins (Ariel and Ryan, 2010; Ariel et al., 2013), and fluorescent membrane proteins engineered to report glutamate release (Helassa et al., 2018; Oertner et al., 2002).

1.2.5. Synaptic vesicle priming and exocytosis

To generate the RRP, a complex molecular machinery is required to organize synaptic vesicles at the presynaptic membrane, including priming and membrane fusion proteins. Munc13 and calcium-dependent secretion activator (CAPS) priming proteins are essential for generating a functionally and molecularly primed pool of fusion-competent vesicles (Augustin et al., 1999; Imig et al., 2014; Jockusch et al., 2007; Varoqueaux et al., 2002). On the molecular level, vesicular priming requires a coordinated interaction with select core components of the exocytotic machinery, namely the neuronal soluble N-ethylmaleimide-factor attachment protein receptors (SNAREs). SNARE proteins comprise the target SNARE (tSNARE) proteins syntaxin-1 and SNAP-25 on the plasma membrane and the vesicular SNARE (vSNARE) synaptobrevin (Bennett et al., 1992; Link et al., 1992; Schiavo et al., 1992; Sollner et al., 1993). Upon a Munc13-mediated switch from a closed to an open conformation, syntaxin-1 interacts with SNAP-25 to preassemble a t-SNARE acceptor complex at active zone release sites (Ma et al., 2011). Upon arrival of a synaptic vesicle, the vSNARE synaptobrevin binds the tSNARE acceptor complex to form a tight, ternary SNARE complex (Fasshauer et al., 1997, 2002; Hatsuzawa et al., 2003). Whereas partial assembly of the SNARE complex is initially sufficient to dock the vesicle to the plasma membrane (Imig et al., 2014), it is the "zippering" together of the vSNARE and tSNAREs to form the final ternary SNARE complex structure which provides enough energy to drive the fusion of the vesicular and plasma membrane lipid bilayers (Hanson et al., 1997; Jahn et al., 2003; Lin and Scheller, 1997; Sollner et al., 1993).

1.2.6. Molecularly primed synaptic vesicles are morphologically docked

Genetic deletion of Munc13 priming proteins (Aravamudan et al., 1999; Augustin et al., 1999; Richmond et al., 1999; Siksou et al., 2009a; Varoqueaux et al., 2002; Weimer et al., 2006) or components of the SNARE complex (Arancillo et al., 2013; Bronk et al., 2007; Schoch et al., 2001; Washbourne et al., 2002) results in absolute loss of or severe deficits in neurotransmission, respectively. Corresponding ultrastructural analyses combining high-pressure cryofixation and electron tomography revealed docking deficits in Schaffer collateral synapses lacking either Munc13s, CAPSs, sytaxin-1, synaptobrevin-2, or SNAP-25 (Imig et al., 2014; Siksou et al., 2009a). As loss of individual SNARE proteins and priming factors severely affect both functional vesicle priming and morphological vesicle docking, the authors concluded that priming and docking are functional and morphological representations of the same process, specifically partial SNARE complex assembly is mediated by Munc13s and CAPSs priming molecules (Imig et al., 2014). Additional converging lines of evidence indicating that functionally primed vesicles are detected at the ultrastructural level as vesicles docked or in close physical contact with the active zone membrane include: (i) the number of docked vesicles correlates closely with the number of vesicles predicted to fuse in response to stimuli triggering the release of all fusion competent vesicles (Schikorski and Stevens, 1997); and (ii) optogenetically and electrically evoked action potentials selectively deplete the docked pool of vesicles (Kusick et al., 2020; Watanabe et al., 2013a).

Moreover, the discovery that synapses from genetic null mutants of Munc13 priming proteins or components of the SNARE complex accumulate vesicles within 5-10 nm of the active zone membrane provided evidence of morphologically distinct steps upstream of the docking/priming process (Imig et al., 2014; Siksou et al., 2009a). These findings appear consistent with the aforementioned data obtained by cryo-electron tomographic analyses of frozen-hydrated synaptosomes, which revealed structural filaments ("tethers") approximately 10 nm in length linking vesicles to the active zone membrane (Fernández-Busnadiego et al., 2010, 2013).

Taken together, these findings indicate that morphologically docked vesicles fulfill at least the molecular requirements to fuse in response to vesicle depleting stimuli, such as hypertonic sucrose. It should however be considered that in response more physiologically relevant stimuli, such as single action potentials, only a subset of these docked and primed vesicles are likely to fuse. Moreover, synapse-specific properties that determine the size of this fusing subset critically determine how the efficacy of transmitter release is dynamically, or plastically, changed during repetitive stimulation.

1.2.7. Release probability and short-term plasticity

Despite the fact that many molecular components of the neurotransmitter release machinery are highly conserved between different synapse types, a remarkable range of functional synaptic behaviors are displayed depending on the synapse type or brain region in question (Südhof, 2012). Two important, and inherently linked, presynaptic parameters that contribute to this functional heterogeneity are synaptic release probability and short-term plasticity.

Synaptic release probability describes the likelihood that a given synaptic vesicle will fuse upon action potential arrival in the presynaptic terminal (Neher, 2015). The closer the release probability is to 1, the more likely it is that a release-ready vesicle fuses during an action potential. Multiple factors can influence the release probability of a synapse, including the number of available fusion-competent vesicles (i.e. the size of the RRP) (Imig et al., 2014; Varoqueaux et al., 2005), the physical distance separating VGCCs from the vesicular calcium sensor (i.e. the coupling distance) (Chen et al., 2015; Rebola et al., 2019; Vyleta and Jonas, 2014), the geometrical arrangement of VGCCs at the active zone membrane (Keller et al., 2015; Miki et al., 2017; Rebola et al., 2019), the type and sensitivity of the vesicular calcium sensor (Chen et al., 2015; Fernández-Chacón et al., 2001; Jackman et al., 2016), and the intrinsic properties of the vesicle related to the state of its release machinery (Cano et al.,

2012). The notion of heterogeneous vesicular release probabilities within the RRP, however, remains controversial (Neher, 2015).

Synapse-specific differences in release probability shape their distinct short-term plasticity characteristics. Short-term plasticity, the alteration of synaptic strength upon repetitive stimulation, was first observed in the form of paired pulse facilitation in the frog neuromuscular junction and paired pulse depression in the cat neuromuscular junction (Eccles et al., 1941). Eccles and colleagues found that the endplate potential, the postsynaptic response in a muscle fiber, in the frog neuromuscular junction increased after two closely spaced stimuli and that this effect diminishes as the interstimulus interval, the time between two subsequent stimuli, increases (Eccles et al., 1941). Conversely, the cat neuromuscular junction undergoes paired-pulse depression in which the second endplate potential is smaller than the first (Eccles et al., 1941). Typically, synapses that undergo paired-pulse facilitation have a low initial release probability, such that more release-ready vesicles remain to fuse in response to subsequent stimuli and associated elevations in presynaptic calcium. Conversely, synapses with higher release probability release a greater proportion of vesicles during the initial stimulus, so that fewer vesicles remain to fuse in response to a second closely-spaced stimulus, thus leading to paired-pulse depression (reviewed in Jackman and Regehr, 2017).

Multiple factors contribute to dynamic changes in synaptic transmission efficacy and thereby to mechanisms mediating short-term plasticity. Short-term plasticity is postulated to be an important mechanism for the forming and processing of memory in the hippocampus (Neves et al., 2008). These factors include: action potential broadening (Geiger and Jonas, 2000); VGCC-vesicular calcium sensor coupling distance (Eggermann et al., 2012; Nakamura et al., 2018; Vyleta and Jonas, 2014); calcium sensors (Fernández-Chacón et al., 2001; Jackman et al., 2016); endogenous presynaptic calcium buffers (Blatow et al., 2003; Dumas et al., 2004; Müller et al., 2005; Vyleta and Jonas, 2014); and the availability of fusion competent synaptic vesicles (Imig et al., 2014; Miki et al., 2020; Siksou et al., 2009a; Südhof and Rizo, 2011).

Calcium influx is altered by a broadening of the action potential spike, which in turn extends the depolarization at the terminal and the number of open VGCCs. Action potential broadening has been studied in the hippocampal mossy fiber synapse where increased activity causes a broadening of the action potential and enhancement of synaptic transmission (Geiger and Jonas, 2000). Geiger and Jonas demonstrated that activity-dependent inactivation of potassium channels causes a broadening of the action potential spike (Geiger and Jonas, 2000). During spike broadening, VGCCs are open for longer periods of time, leading to increased presynaptic calcium concentrations and a concomitant enhancement of synaptic vesicle fusion (Geiger and Jonas, 2000).

Although the length of time a calcium channel is open during an action potential can modulate synaptic transmission, the distance between calcium channels and sensors located on synaptic vesicles is also important for synaptic efficacy. By loading presynaptic boutons with calcium chelators with different binding kinetics, the distance of synaptic vesicles to VGCCs can be estimated based on the degree synaptic transmission is reduced (Chen et al., 2015; Eggermann et al., 2012; Vyleta and Jonas, 2014). For example, the coupling distance of fast-releasing synaptic vesicles in the calyx of Held has been estimated to be approximately 16 nm (Chen et al., 2015), whereas a looser coupling of approximately 70 nm has been postulated for hippocampal mossy fiber synapses (Vyleta and Jonas, 2014). In another study, coupling distance was attributed to release probability in cerebellar stellate and granule cells, in which neurons with higher release probability had a tighter coupling distance than facilitating neurons (Rebola et al., 2019).

Another mechanism of facilitation is buffer saturation (Jackman and Regehr, 2017). Buffer saturation is caused by endogenous calcium-binding molecules buffering out free calcium ions upon calcium influx at the start of repeated stimulation (Jackman and Regehr, 2017). The remaining few free calcium ions trigger the fusion of few synaptic vesicles. Upon the arrival of a subsequent action potential, the endogenous buffers remain saturated, and more free calcium is available to trigger fusion of more synaptic vesicles (Jackman and Regehr, 2017). For the buffer saturation model to contribute to synaptic facilitation, a combination of high-affinity calcium buffers, high concentrations of calcium buffers, and relatively large distances between the calcium channels and sensors is required (Jackman and Regehr, 2017). For example, hippocampal mossy fiber synapses have a fast-acting calcium buffer, calbindin, with a high affinity for calcium that can rapidly buffer free calcium during a single action potential (Eggermann et al., 2012; Nägerl et al., 2000). Consequently, mossy fiber synapses exhibit short-term facilitation and a low release probability (Kawamura et al., 2004; Salin et al., 1996; Toth et al., 2000; Vyleta and Jonas, 2014). Other endogenous calcium buffers, such as calmodulin, are also found in mossy fiber boutons (Chamberland et al., 2018; Salin et al., 1996; Xia et al., 1991).

The type of calcium sensor residing on synaptic vesicles, such as synaptotagmins (Craxton, 2010; Südhof, 2002), may also contribute to short-term facilitation. While synaptotagmin-1 and synaptotagmin-2 isoforms are well-known calcium sensors for synaptic vesicle fusion, they contribute primarily to the fast component of transmitter release, likely on the first action potential during a series of action potentials (Brandt et al., 2012; Hui et al., 2005). Another calcium sensor, synaptotagmin-7, has very high calcium affinity, but slow disassociation kinetics as determined through *in vitro* studies (Brandt et al., 2012; Hui et al., 2005) and has been shown to contribute to facilitation in Schaffer collateral and mossy fiber synapses (Jackman et al., 2016). The presence of an additional calcium sensor with properties similar to synaptotagmin-7 could be a later-phase or asynchronous sensor involved in synaptic vesicle fusion in the second, facilitating pulse (Hui et al., 2005).

Lastly, and of particular importance to the motivation to perform this study, the availability of fusion-competent, docked synaptic vesicles may contribute to the functional heterogeneity of synapses (Dobrunz, 2002; Dobrunz and Stevens, 1997; Schikorski and Stevens, 1997). Based on the assumption that morphologically docked vesicles overlap with the functional RRP (Imig et al., 2014; Kusick et al., 2020; Schikorski and Stevens, 1997; Siksou et al., 2009a; Watanabe et al., 2013a) many previous studies have attempted to elucidate whether the availability of docked synaptic vesicles contribute to synapse-specific differences in synaptic release probability and plasticity characteristic (Eltes et al., 2017; Holderith et al., 2012; Millar et al., 2002; Xu-Friedman et al., 2001). These studies failed to come to a strong consensus, possibly due to the variety of organisms, brain regions and synapse types examined, and variations in the methodological approaches used.

As an example, a comparative analysis of climbing fiber (low release probability, facilitating) and parallel fiber (high release probability; depressing) synapses in the cerebellum of perfusion-fixed rats found no difference in the availability of docked synaptic vesicles analyzed by three-dimensional (3D) serial section EM (Xu-Friedman et al., 2001). A study of associational/commissural synapses onto *cornu ammonis* area 3 (CA3) pyramidal neurons in the hippocampus of chemically-fixed acute rat slices found that synapses with low release probability had fewer docked synaptic vesicles than those with high release probability analyzed by 3D serial EM (Holderith et al., 2012). However, excitatory CA3 pyramidal synapses onto metabotropic glutamate receptor type1α-positive interneurons (facilitating synapses) harbored fewer docked vesicles than synapses onto parvalbumin-positive interneurons (depressing synapses) in the hippocampus of perfusion-fixed rats analyzed by electron tomography (Eltes et al., 2017). Conversely, a comparative ultrastructural study of phasic (depressing) and tonic (facilitating) motor neurons found tonic motor neurons harbored more docked synaptic vesicles than phasic motor neurons of the main leg extensor muscle of freshwater crayfish chemically fixed and analyzed by 3D serial EM (Millar et al., 2002). Taken together, these studies do not come to a consensus regarding the number of morphologically available synaptic vesicles in shaping synaptic functional properties.

To understand these conflicting findings, I address the same question in this book using state-of-the-art methodological approaches introduced later in this section to perform a comparative ultrastructural analysis of two extensively characterized excitatory synapses in the hippocampal formation, mossy fiber-CA3 and Schaffer collateral synapses. As indicated above, many factors have been implicated as mechanisms contributing to mossy fiber short-term facilitation (Chamberland et al., 2014; Dumas et al., 2004; Geiger and Jonas, 2000; Jackman et al., 2016; Vyleta and Jonas, 2014). However, the relationship between the organization of synaptic vesicles and the plasticity characteristics of this synapse remains unclear. This information is ultimately required to fully understand mossy fiber synapse function, both at the level of the synapse and in the context of the hippocampal network.

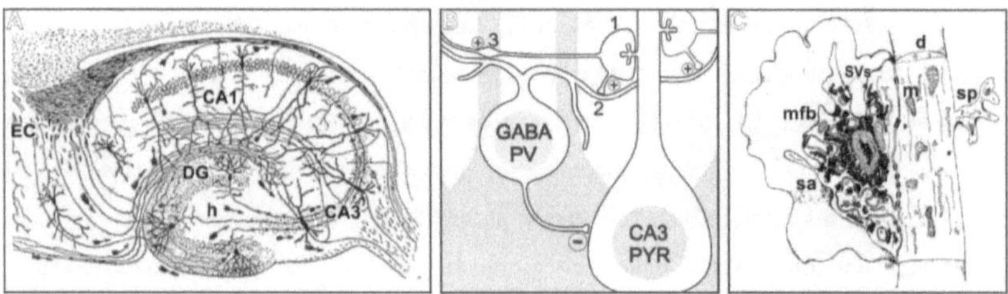

Figure 3. Hippocampal network connectivity and the mossy fiber pathway.

(**A**) Connectivity of neurons in the hippocampus as originally described by Ramon y Cajal. Granule cells with cell bodies located in the dentate gyrus (DG) send axonal projections to the CA3 where they form three types of synaptic connections. (**B**) Illustration of synaptic connections formed by granule cell mossy fibers in the hippocampus. Mossy fibers form 1) mossy fiber boutons onto the primary dendrite and thorny spine excrescences of CA3 pyramidal neurons; 2) filopodial extensions from the mossy fiber boutons; and 3) axonal *en passant* boutons onto inhibitory interneurons that feedforward onto CA3 pyramidal neurons. (+) and (-) symbols represent excitatory glutamatergic and inhibitory GABAergic synapses, respectively. (**C**) Depiction of a mossy fiber bouton (mfb) and complex thorny spine excrescences (sp) of CA3 pyramidal neurons adapted from Amaral and Dent, 1981. Abbreviations: CA1/CA3, *cornu ammonis* area 1 and 3; DG, dentate gyrus; EC, entorhinal cortex; h, hilus; d, dendrite; m, mitochondria; mfb, mossy fiber bouton; sa, spine apparatus; sp, thorny spine excrescence; SVs, synaptic vesicles. *Permission & Rights: (A) Adapted from* Nicoll and Schmitz, 2005 *through Copyright Clearance Center with license number 4786480011593*. (**C**) *Adapted from* Amaral and Dent, 1981 *through Copyright Clearance Center with license number 4786540090843*.

1.3. The role of functional heterogeneity in brain circuits: the hippocampus

The hippocampal formation is the brain area in which information from the cortex is processed for storage in the form of memory (Jarrard, 1993). Memory formation relies on a well-described anatomical circuit of excitatory neural connections called the tri-synaptic pathway (Andersen et al., 1966). The tri-synaptic pathway is composed of three different pathways, the perforant, the mossy fiber, and the Schaffer collateral pathways. The perforant pathway comprises axonal projections from the entorhinal cortex (EC) that form excitatory synapses onto the granule cells of the dentate gyrus (DG) (Figure 3 A) (Blackstad and Kjaerheim, 1961). The second synaptic connection is the mossy fiber pathway where granule cells send axonal projections to the CA3 where they form giant, excitatory mossy fiber synapses onto CA3 pyramidal neurons (Figure 3 A) (Blackstad and Kjaerheim, 1961). Finally, in the Schaffer collateral pathway, CA3 pyramidal neurons send axonal projections to the CA1 where they form synaptic connections with pyramidal neurons (Figure 3 A) (Blackstad and Kjaerheim, 1961). Outgoing axonal projections run from the CA1 to layer V neurons in the EC via subiculum (Ceccom et al., 2014).

1.3.1. Mossy fiber pathway

Although a simplistic anatomical view is helpful to understand the "flow" of information through the hippocampus, a deeper consideration of local circuitry, including feedforward and feedback loops, and of synapse-specific functional properties and plasticity characteristics is ultimately required to understand or predict the behavior of complex neuronal networks. The hippocampal mossy fiber projection serves as an excellent example of how such complexity shapes its role within the tri-synaptic pathway.

Mossy fiber synapses are generated from granule cell axonal projections to the *stratum lucidum* of the CA3 and establish excitatory synaptic connections with pyramidal neurons along the first 100 μm of the primary dendrites via large *en passant* boutons (Henze et al., 2000). One granule cell forms 15-18 large mossy fiber boutons (Amaral and Dent, 1981) and each pyramidal neuron in the CA3 is innervated by approximately 50 mossy fiber boutons (Amaral et al., 1990). Demonstrative of the mossy fiber projection's unique level of target specificity, small granule cell axonal varicosities and filopodial extensions emerging from the bouton form synapses onto local inhibitory interneurons that target the soma of CA3 pyramidal cells (Figure 3 B). In total, each granule cell axon forms approximately 150 synapses onto inhibitory interneurons. Whereas large synapses formed by large mossy fiber boutons exhibit very low initial release probability and profound frequency facilitation (Lawrence et al., 2004; Salin et al., 1996), synapses formed by small varicosities and filopodial extensions depress in during repetitive firing (Toth et al., 2000). Based on this combination of anatomical and physiological properties, the net effect of granule cell firing at basal frequencies is one of feedforward inhibition of CA3 pyramidal cells (Acsády et al., 1998). In contrast, during elevated firing rates, the profound frequency facilitation (Salin et al., 1996) exhibited by mossy fiber-CA3 synapses overcomes this feedforward inhibition and fulfills a "conditional detonator" function with a profound influence on the excitability of postsynaptic CA3 pyramidal cells (Henze et al., 2002a).

1.3.2. Structural comparison

The low initial release probability of the mossy fiber-CA3 synapse is perhaps unexpected considering ultrastructural features including the total number of synaptic vesicles and active zone release sites harbored within individual giant mossy fiber boutons (Amaral and Dent, 1981; Chicurel and Harris, 1992; Henze et al., 2000; Rollenhagen et al., 2007; Wilke et al.,

2013). Mossy fiber boutons are several microns in diameter and filled with tens of thousands (~20,500) of synaptic vesicles (Figure 3 C; Table 1; Chicurel and Harris, 1992; Rollenhagen et al., 2007). These boutons are easily distinguishable in ultrastructural studies due to their large size as well as the presence of multiple synaptic contacts onto the CA3 pyramidal neuron (Table 1; Chicurel and Harris, 1992; Rollenhagen et al., 2007). Mossy fiber boutons form synapses onto specialized, multi-headed spines on CA3 pyramidal neurons called thorny excrescences (Figure 3 C; Table 1; Chicurel and Harris, 1992). Unlike classically described dendritic spines, thorny excrescences tend to contain many organelles, including microtubules, multivesicular bodies, and spine apparatuses (Chicurel and Harris, 1992).

Presynaptically, giant mossy fiber boutons are characterized by a relative abundance of mitochondria, microtubules, and mutivesicular bodies (Amaral and Dent, 1981; Chicurel and Harris, 1992; Rollenhagen et al., 2007). In addition to the small clear-core vesicles typical of glutamatergic synapses, mossy fiber synapses also harbor clear-core synaptic vesicles of considerably larger dimensions (Henze et al., 2002b; Laatsch and Cowan, 1966) and dense-core vesicles (DCVs) (Amaral and Dent, 1981; Rollenhagen et al., 2007). DCVs are characterized in electron micrographs as large vesicles (70-100 nm in diameter) with electron-dense cores (Amaral and Dent, 1981).

The origin and functional implications of giant clear-core vesicles remain largely unknown, although it has been previously postulated that they contribute to glutamatergic signaling between granule cells and postsynaptic CA3 pyramidal cells (Henze et al., 1997). Based on the

Table 1. Comparative morphologies of Schaffer collateral and mossy fiber synapses from ultrastructural studies.

	Averages	Schaffer collateral synapses	Mossy fiber synapses
Presynapse	Bouton volume (μm^3)	0.11[1]	7-8[2]
	Number of SVs	223[1]	~20,400[2]
	Number of AZs per bouton	1[1]	29.75[2]
	AZ area (μm^2)	~0.03[3]	0.12[2]
Postsynapse	Spine volume (μm^3)	0.03[1]	0.13-1.83[2]
	Spine area (μm^2)	0.63[1]	16-23[2]
	Total PSD area (μm^2)	0.069[1]	1-3[2]

Abbreviations: AZ, active zone, PSD, postsynaptic density; SVs, synaptic vesicles.
[1]Harris and Stevens, 1989; [2] Rollenhagen et al., 2007; [3] Shepherd and Harris, 1998

large lumenal volume (i.e. potential neurotransmitter capacity) and the proximity of giant mossy fiber boutons to the cell bodies of CA3 pyramidal cells, the fusion of such large vesicles in this scenario could be expected to profoundly influence the excitability of the postsynaptic neuron. Despite this interesting hypothesis, giant vesicles remain enigmatic and many important questions are still to be addressed. For example: i) Do giant vesicles dock in physical contact with active zone membranes and are they molecularly equipped to be primed for fusion? ii) Through what mechanisms are giant vesicles formed and are they regulated by synaptic activity?

Although the mossy fiber pathway uses glutamate as a primary neurotransmitter, other messengers such as neuropeptides (Chavkin et al., 1983; Henze et al., 2000; McQuiston and Colmers, 1996; Salin et al., 1995; Sherwood and Lo, 1999; Weisskopf et al., 1993), zinc (Lavoie et al., 2011), and adenosine triphosphate (ATP) (Henze et al., 2000; Loewen et al., 1992; Yamamoto et al., 1993), modulate synaptic function. Neuropeptides produced in hippocampal granule cells and trafficked to mossy fiber boutons (Henze et al., 2000), include, but are not limited to, brain-derived neurotrophic factor (BDNF), enkephalins, dynorphin, and neuropeptide Y (Chavkin et al., 1983; Henze et al., 2000) These relatively large peptide signaling molecules (3-36 amino acids; compared to single amino acids like glutamate and GABA) are synthesized in the rough endoplasmic reticulum within the cell soma, and transported to the Golgi apparatus (Gondré-Lewis et al., 2012). Once at the Golgi apparatus, they are packaged into DCVs (Commons and Milner, 1995; Dieni et al., 2012, 2015) and transported along microtubules via anterograde axonal transport to presynaptic boutons (Gondré-Lewis et al., 2012). Upon DCV fusion, neuropeptides can affect synaptic transmission via actions on both presynaptic and postsynaptic targets (Chavkin et al., 1983; McQuiston and Colmers, 1996; Salin et al., 1995; Sherwood and Lo, 1999; Weisskopf et al., 1993).

In contrast to the intriguing morphology of hippocampal mossy fiber synapses, Schaffer collateral synapses represent prototypical examples of small, glutamatergic synapses. On the ultrastructural level, Schaffer collateral synapses are visualized as small axonal varicosities filled with several hundred synaptic vesicles (Harris and Stevens, 1989) that cluster at asymmetric synaptic contacts onto postsynaptic dendritic spines of CA1 pyramidal cells in the *stratum radiatum* (Figure 1 B; Table 1; Gray, 1959; Harris and Stevens, 1989). Schaffer collateral synapses typically form one synapse onto one spine, however, on occasion one presynaptic bouton can contact multiple spines (Table 1; Harris and Stevens, 1989). Other organelles, such as DCVs, microtubules, and mitochondria are also present in the Schaffer collateral presynaptic compartment (Harris and Stevens, 1989). The postsynaptic spine lacks organelles with the exception of the occasional spine apparatus (appearing in approximately 23% of mature CA1 spines), a lamellar stack of smooth-endoplasmic reticulum membranes with electron-dense F-actin filaments between the folds of each lamella (Capani et al., 2001; Spacek and Harris, 1997).

1.4. Schaffer collateral and mossy fiber synapses

1.4.1. Functional differences between synapses

The two synapses I focus on in this study have well-characterized differences in synaptic efficacy. Both synapse types exhibit paired pulse facilitation indicating a low release probability. However, the release probability of Schaffer collateral synapses is heterogeneous in acute slices when using an external fluorescent indicator that binds glutamate (Helassa et al., 2018; Oertner et al., 2002), and is considerably higher than in mossy fiber synapses (Lawrence et al., 2004). For comparison, Schaffer collateral synapses facilitate for the first few action potentials and then depress during high frequency stimulation (Dobrunz and Stevens, 1997, 1999; Helassa et al., 2018; Jackman et al., 2016; Oertner et al., 2002). In comparison, mossy fiber synapses exhibit strong frequency facilitation (Table 2; Jackman et al., 2016; Salin et al., 1996). Frequency facilitation is the reversible enhancement of synaptic transmission at low frequency stimulation (Salin et al., 1996). For example, the amplitude of mossy fiber-driven EPSCs increases by up to 300% of baseline when the frequency of the stimulus changes from 0.05 Hz to 0.2 Hz in acute guinea pig hippocampal slices (Salin et al., 1996).

Synaptic ultrastructure-function relationships can also be probed in individual synapses by modulating their functional state. Various pharmacological manipulations that target specific presynaptic molecules have been utilized in past studies to induce changes in neurotransmitter release efficacy. For example, mossy fiber synaptic transmission is strongly potentiated following the bath application of either tetraethylammonium (TEA) (Suzuki and Okada, 2008; Zhao et al., 2012a) or forskolin (Weisskopf et al., 1994). Although TEA and forskolin both act via presynaptic mechanisms to induce a prolonged state of enhanced mossy

Table 2. Functional properties of Schaffer collateral and mossy fiber synapses.

Averages	Schaffer collateral synapses	Mossy fiber synapses
Readily releasable pool of vesicles	~10 per synapse[1]	400-1400 per bouton[5,6]
Morphological RRP estimates (per AZ)	10-12[2]	36.6[5]
Release probability	0.2-0.6[3]	<0.1[7]
Short-term plasticity characteristics	Mild-facilitation[4]	Strong frequency facilitation[8]

[1]Stevens and Tsujimoto, 1995; [2]Imig et al., 2014; [3]Oertner et al., 2002; [4]Jackman et al., 2016; [5]Rollenhagen et al., 2007; [6]Midorikawa and Sakaba, 2017; [7]Vyleta and Jonas, 2014; [8]Salin et al., 1996. Abbreviation: AZ, active zone.

fiber transmission referred to as chemical long-term potentiation (LTP), the two drugs target different molecules. Whereas TEA blocks presynaptic potassium channels, thereby broadening the action potential spike, prolonging VGCC channel opening, and increasing the presynaptic influx of calcium (Ishikawa et al., 2003; Suzuki and Okada, 2008), forskolin activates adenylate cyclase 1 (AC1), which enhances mossy fiber synaptic transmission via increased presynaptic cyclic adenosine monophosphate (cAMP) (Barovsky et al., 1984; Weisskopf et al., 1994). When calcium enters the presynaptic terminal via VGCCs, it triggers synaptic vesicle fusion and also acts as a second messenger to activate AC1 (Villacres et al., 1998), which converts ATP into cAMP (Chavez-Noriega and Stevens, 1994). Although the presynaptic signaling cascades and molecular mechanisms underpinning cAMP-dependent enhancement of mossy fiber synaptic transmission have been partially elucidated (Henze et al., 2000; Huang et al., 1994a; De Rooij et al., 1998; Trudeau et al., 1996; Tzounopoulos et al., 1998; Villacres et al., 1998; Weisskopf et al., 1994), it remains unclear precisely how they ultimately manifest as changes in release probability at active zone release sites. For example, both the enzyme protein kinase A (PKA) (Weisskopf et al., 1994) and exchange protein directly activated by cAMP (Epac) (Fernandes et al., 2015) have been implicated as downstream targets of cAMP signaling involved in enhancing presynaptic mossy fiber release probability. Although several candidate molecules have been postulated to act as downstream effectors in these pathways (Castillo et al., 2002; Cho et al., 2015; Fykse et al., 1995; Kaeser-Woo et al., 2013; Lonart and Sudhof, 1998), their potential influence on the structural organization of active zone release sites remains to be investigated in the context of enhanced transmitter release probability.

Application of the metabotropic glutamate receptor type 2 (mGluR2) agonist (2S,1'R,2'R,3'R)-2-(2,3-dicarboxy-cyclopropyl)glycine (DCG-IV), decreases mossy fiber presynaptic cAMP levels (Chen et al., 2001; Huang et al., 2002). DCG-IV acts upon presynaptically expressed mGluR2 receptors, causing a decrease in evoked and spontaneous release of synaptic vesicles at mossy fiber synapses (Kamiya and Ozawa, 1999; Kamiya et al., 1996). Consequently, DCG-IV is commonly used in electrophysiological experiments to demonstrate that EPSCs recorded from CA3 pyramidal cells are of mossy fiber origin (Breustedt et al., 2010; Brockmann et al., 2019; Castillo et al., 2002; Galimberti et al., 2006; Kamiya et al., 1996; Nicoll and Schmitz, 2005). In the hippocampus, mGluR2 expression is specific to the presynaptic terminals of mossy fiber synapses (Ikeda et al., 1995). DCG-IV suppresses synaptic transmission by reducing presynaptic calcium influx as well as by acting on the vesicular release machinery, however the mechanism of action is not fully understood (Kamiya and Ozawa, 1999).

In summary, the functional state of hippocampal mossy fiber synapses is highly sensitive to presynaptic cAMP levels (Fernandes et al., 2015; Villacres et al., 1998; Weisskopf et al., 1994), which can be bidirectionally manipulated by forskolin or DCG-IV treatment. My objective is to exploit these pharmacologically induced changes in mossy fiber transmission efficacy to examine a potential link between cAMP-dependent alterations in release probability and changes in the availability of morphologically docked synaptic vesicles at active zone release sites. This has not previously been investigated in hippocampal mossy fiber synapses.

1.4.2. Correlating structure and function

The main objective of this study is to examine the relationship between the availability of morphologically docked synaptic vesicles, thus the RRP, and release probability and synaptic strength. My decision to focus primarily on the hippocampal mossy fiber synapses was motivated by several factors. Most importantly, the mossy fiber-CA3 synapses is characterized by a very low release probability and strong frequency facilitation (Lawrence et al., 2004; Salin et al., 1996; Vyleta and Jonas, 2014; Vyleta et al., 2016). Moreover, both short- and long-term plasticity at mossy fiber synapses are regulated presynaptically (Salin et al., 1996; Toth et al., 2000; Zalutsky and Nicoll, 1990). Whereas the anatomical connectivity and gross presynaptic morphology of the mossy fiber-CA3 projection has been well characterized (Acsády et al., 1998; Chicurel and Harris, 1992; Rollenhagen et al., 2007; Sai et al., 2017; Wilke et al., 2013), the fine structural organization of mossy fiber active zones has not previously been scrutinized at a resolution permitting accurate discrimination of functionally distinct vesicle pools. Consequently, the relative contribution of active zone structural organization cannot be assessed in parallel with other postulated mechanisms (e.g. coupling distance, calcium buffer species, and action potential broadening) of mossy fiber release probability. Hippocampal mossy fibers are also of particular advantage for relating ultrastructural observations with recorded functional parameters. The large size of presynaptic mossy fiber boutons permits direct presynaptic capacitance recordings to be made during step-depolarizations (Hallermann et al., 2003; Midorikawa and Sakaba, 2017). Thus, quantitative morphological estimates of docked and membrane-proximal vesicle pools obtained by electron tomography can be compared with functional RRP estimates (Hallermann et al., 2003; Midorikawa and Sakaba, 2017). Based on these factors, I was motivated to revisit the question of whether the functional status of a synapse can be predicted by the spatial organization of membrane proximal vesicles. My hypothesis is that an experimental approach combining organotypic hippocampal slice culture, rapid high-pressure freezing (HPF) cryo-fixation, and 3D electron tomography will provide a novel, and more accurate, perspective of synaptic ultrastructure-function relationships by circumventing methodological limitations associated with past studies (see Discussion section *Methodological considerations*).

1.5. Methodology

1.5.1 Hippocampal slice cultures

Most studies of synaptic and network function rely on access to brain tissue either maintained *in vitro* or acutely prepared slices of isolated brain regions. Exhibiting the least complex neuronal circuitry, autaptically cultured neurons are isolated and grown on astrocyte islands, forcing them to makes synapses onto themselves (Bekkers and Stevens, 1991; Millet and Gillette, 2012). In contrast, continental, or dissociated, primary neuron cultures generated from dissociated brain tissue establish synaptic connections with one another in a culture dish (Harrison, 1910; Millet and Gillette, 2012). Although both these reduced complexity culture systems are advantageous for assessing fundamental functional synaptic properties, not all synapse types can be recapitulated in cultured monolayers and many plasticity characteristics exhibited by synapses *in vivo* are not comparably expressed *in vitro* (Mennerick and Zorumski, 1995).

Organotypic slice culture systems maintain a near-native network connectivity, such thereby circumventing, or greatly mitigating many of the aforementioned limitations (Debanne et al., 1996; Gähwiler, 1984; Galimberti et al., 2006; Humpel, 2016; Marchal and Mulle, 2004). Moreover, cultures can be used to maintain neural networks in animals with post-natal lethality and to identify specific synapses within the tissue due to stereotyped targeting within the circuit (Imig et al., 2014). Slices cultured according to the interface method are maintained at the gas-liquid interface on top of a semi-permeable membrane insert, which provides access to slice culture medium containing metabolic nutrients (Stoppini et al., 1991). Hippocampal slice cultures have been extensively used in ultrastructural studies (Galimberti et al., 2006; Imig and Cooper, 2017; Studer et al., 2014; Zhao et al., 2012b, 2012c, 2012a) since they contain synaptic networks with well-studied synaptic functional characteristics that differ from one another (Eltes et al., 2017; Holderith et al., 2012; Jackman et al., 2016; Nicoll and Schmitz, 2005; Salin et al., 1996). As the slice recovers from the trauma of sectioning, minor rearrangements of the can tissue occur, however the stereotypical cellular lamination of the hippocampus remains intact (Buchs et al., 1993). Possible changes in the network connectivity include an increase in synapse density in the *stratum ratiatum* of the CA1 as a compensatory mechanism due to the loss of other synaptic inputs from other brain areas (Muller et al., 1993). Granule cells have been found to sprout in hippocampal slice cultures and form excitatory inputs back to the granule cell layer (Coltman et al., 1995; Frotscher et al., 2006), however simply culturing more EC remedies sprouting (Coltman et al., 1995).

1.5.2. Fixation methods

1.5.2.1. Aldehyde fixation

Conventional tissue fixation methods have long included the use of aldehydes. Two types of aldehydes are frequently used for tissue fixation and ultrastructural studies. Paraformaldehyde (PFA) and glutaraldehyde (GA) are the most commonly used aldehydes, which fix tissue by cross-linking free amino acids (Bullock, 1984; Hopwood, 1969). Several structural artifacts have been attributed to aldehyde fixation protocols for EM, including deformation of subsynaptic organelles (Murk et al., 2003) and depletion of membrane-proximal synaptic vesicle pools (Korogod et al., 2015; Smith and Reese, 1980). However, aldehydes can be used to fix large tissue volumes and has been particularly advantageous for large-scale ultrastructural studies (Korogod et al., 2015).

1.5.2.2. High-pressure freezing

A method called HPF (high-pressure freezing) uses the principle of super-cooling water under high pressures to cause water vitrification rather than ice crystallization and can improve the penetration depth of cryopreservation in biological tissues (Dahl and Staehelin, 1989; Kanno et al., 1975; Moor, 1987; Moor et al., 1980). HPF creates high pressure and rapid cooling at the same time by forcing liquid nitrogen on the top and bottom of a sample at a high velocity. To achieve ideal cryopreservation, pressure on the sample must reach 2000 bar before the temperature drops below 0° C (Dahl and Staehelin, 1989). If the timing of either parameter is off, this can lead to poor ultrastructural preservation in the form of either ice crystal damage, as in instances when temperature reached 0° C before pressure reaches 2000 bar, or pressure artifacts, as in instances when pressure increases before the temperature drops (Möbius et al., 2010). Both artifacts can only be assessed at the electron microscopic level. Potential caveats of HPF include, (i) sample dimensions limited to a size compatible with the carriers used for HPF (Dahl and Staehelin, 1989; Korogod et al., 2015), and (ii) the depth of ideal tissue cryopreservation depends on the water content of the sample which can be partially mitigated by the use of additional external cryoprotectants (Dahl and Staehelin, 1989).

Fixation of tissue with HPF coupled with automated freeze substitution (AFS) can capture dynamic cellular processes in near-native conditions with little to no alteration to tissue ultrastructure. AFS is the process in which tissue can be infiltrated with solvents for ideal plastic polymers at very low temperatures to prevent unwanted alterations to cellular ultrastructure (Giddings, 2003). Once the tissue has been dehydrated and contrasted with heavy metal solutions such as osmium tetroxide (OsO_4) then the tissue can be infiltrated with plastic polymers such as Epon to embed for EM (Finck, 1960; Giddings, 2003).

1.5.3. Electron microscopy

1.5.3.1. Transmission electron microscopy

Transmission electron microscopes (TEMs) are an ideal way to investigate nano-scale structures in tissue samples. TEMs pass a beam of electrons perpendicular to and directly through a plastic-embedded sample to generate a magnified image in a viewing screen or camera below. In the field of neuroscience, EM has allowed many important ultrastructural observations of tissue samples. Transmission EM has been a key method in linking functional studies to ultrastructure. Typically, ultrathin sections 50-60 nm thick have been used to assess vesicle clustering in proximity to and docking at the presynaptic membrane. Membrane curvature, overlapping synaptic vesicle projections, and incomplete cross-sections of synaptic vesicles can all confound the accurate detection of docked and tethered synaptic vesicles in two-dimensional (2D) EM analyses (Imig and Cooper, 2017; Verhage and Sørensen, 2008). Limited z-resolution of 2D transmission EM can lead to an inaccurate assessment of ultrastructural measurements such as an average vesicle diameter of 33-36 nm from 2D ultrastructural studies (Harris and Sultan, 1995) versus ~45 nm reported via 3D electron tomography (Imig et al., 2014). While the resolution in x and y dimensions can achieve resolutions less than 1 nm, the z-resolution of transmission EM of ultrathin section is restricted by the mechanical limitations of ultramicrotomy (thinnest sections approximately 20 nm-thick; Holderith et al., 2012). Furthermore, the accurate assessment of synaptic vesicle

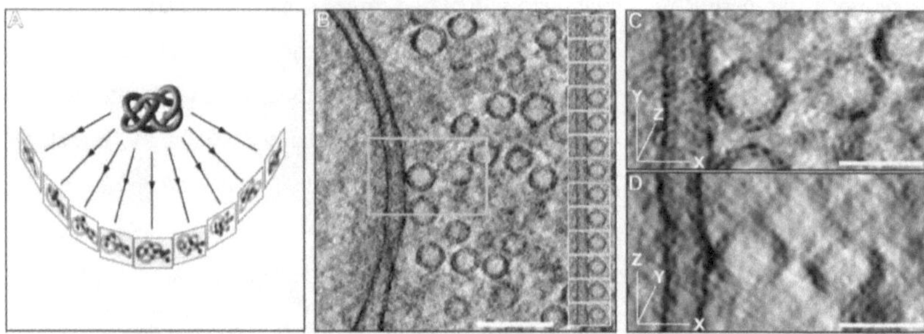

Figure 4. Electron tomography.

(A) Electron tomography works under the principle of reconstructing a three-dimensional object by taking a series of 2D images at many angles. For the electron tomography used in this study, specimens were tilted from -60° to +60° and 2D images acquired every 1°. Using a weighted pack-projection algorithm the series of 2D images are reconstructed into tomograms. (B) Slicer image of a reconstructed synaptic subvolume imaged at 30,000x magnification and binned 3x during reconstruction of the tomogram yielded isotropic voxel dimensions of x, y, and z = 1.6 nm. Docked synaptic vesicles (yellow box and inset) could be accurately assessed in tomograms my moving through the z-stack to determine whether there was no measurable distance between synaptic vesicle and plasma membranes (inset: 13 consecutive sections of the reconstructed subvolume). (C-D) Higher magnification of docked synaptic vesicle from B in x, y (C) and z, x (D) showing that this vesicle definitively fits the docking criteria set for this study. Scale Bars: 100 nm, B; 50 nm, C-D. Permission & Rights: (A) *Adapted from Lučić et al., 2005 with permission through Copyright Clearance Center with order number 1051526.* (B-D) *Adapted from Imig and Cooper, 2017 with permission through Copyright Clearance Center with license number 4877680012945.*

docking requires a better z-resolution that can be achieved with electron tomography (Fernández-Busnadiego et al., 2010, 2013; Imig and Cooper, 2017; Siksou et al., 2009a, 2009b; Stigloher et al., 2011; Vogl et al., 2015).

1.5.3.2. Electron tomography

The theoretical idea of using a set of projection images to generate a 3D reconstructed image was first proposed in the 1960s (Baumeister, 2002; De Rosier and Klug, 1968). Electron tomography works by creating projection images of a specimen, in the case of the present study, tissue embedded in plastic (Figure 4 A) (Baumeister, 2002). Projection images are created by tilting the specimen around an axis that sits perpendicular to the electron beam and taking 2D images incrementally through the entire tilt series (Figure 4 A) (Baumeister, 2002). In my study, specimens were tilted from -60° to +60° and 2D images were taken every one degree (Figure 4; Kremer et al., 1996; Mastronarde, 2005). Raw image files are aligned and a weighted back-projection algorithm calculates a density map in order to reconstruct the final 3D tomogram (Figure 4 B) (Baumeister, 2002; Kremer et al., 1996; Mastronarde,

2005). Limitations of electron tomography include the thickness of the section and the strength of the electron beam generated in the microscope (Baumeister, 2002). High-voltage electron microscopes create an electron beam powerful enough to penetrate the thicker sections required for specimens at high tilt angles. For example, at a tilt angle of 70°, a 200 nm-thick section (at 0°) will be over 500 nm-thick (Baumeister, 2002). At these extreme tilt angles, considerable scattering of the electron beam occurs and image quality drops (Baumeister, 2002). Depending on the imaging magnification, section thickness, camera used for image acquisition, and degree of image binning during tomogram reconstruction, the voxel dimensions and thus the resolution will vary. I achieved a voxel dimension of 1.554 nm by imaging 200 nm-thick sections at 30,000x magnification and binning by 3 during tomogram reconstruction (see Materials and Methods; Figure 4 B-D). With electron tomography, I could more accurately measure synaptic vesicle diameters by finding the exact midpoint of a given synaptic vesicle and precisely measuring the distance between synaptic vesicles and the presynaptic membrane. Furthermore, tomograms can better reveal if a vesicle is morphologically docked or in a membrane proximal position, such as tethered synaptic vesicles within the synaptic subvolume (Figure 4 C and D).

1.6. Purpose of this study

1. There is agreement in the field that the number of morphologically docked synaptic vesicles correlates with the RRP and that docking and priming are morphological and functional manifestations of the same process, however the link between release probability and morphologically docked synaptic vesicles is poorly understood. Therefore, I used state-of-the-art cryopreservation of unstimulated hippocampal slice cultures from wild-type mice to probe whether Schaffer collateral and mossy fiber synapses organize vesicle pools in ways that shape the functional properties at each synapse, i.e. short-term plasticity, RRP, and release probability.

2. Mossy fiber synapses contain a population of large, clear-core vesicles with diameters greater than 60 nm (giant vesicles) whose purpose, molecular identity, and function are poorly-understood (Henze et al., 2002b; Laatsch and Cowan, 1966). These are postulated to be the morphological correlates of giant monoquantal events in rodent acute slices observed in CA3 pyramidal neurons and are of mossy fiber origin (Henze et al., 1997, 2002b). I explore changes in giant vesicle populations in *ex vivo* and *in vitro* preparations, after pharmacological blockade of network activity, and in mossy fiber synapses lacking Munc13 priming proteins to determine if giant vesicles are activity-dependent or an artifact of slice cultures. Further, I explore whether giant monoquantal events occur in mossy fiber synapses from hippocampal slice culture and whether the morphological and functional observations are correlated.

3. Mossy fiber synapses contain a number of DCVs and secrete a variety of neuropeptides (Commons and Milner, 1995; Danzer and McNamara, 2004; Dieni et al., 2015; Henze et al., 2000; McQuiston and Colmers, 1996; Rollenhagen et al., 2007; Salin et al., 1995), however the exact location of DCV exocytosis and whether DCVs fuse under basal conditions in mossy fiber boutons are unknown. I examine changes in the spatial distribution of DCVs in mossy fiber synapses at rest from *ex vivo* and *in vitro* tissue preparations. Furthermore, I explore the effects of pharmacological manipulation of presynaptic cAMP levels on the spatial distribution of DCVs in mossy fiber boutons since neuropeptide secretion in mossy fibers is implicated in mossy fiber synaptic plasticity (Li et al., 2010; Nakamura et al., 2007; Salin et al., 1995).

2. Materials and Methods

The following methods section was written by me and has been published in Cell Reports (Maus et al., 2020).

2.1. Hippocampal slice cultures:

Table 3. Reagents and equipment for hippocampal organotypic slice cultures.

Reagents for hippocampal slice culture	Source	Catalog Number
Basal Medium Eagle (BME)	Thermo Fisher Scientific	Cat No. 41010026
GlutaMAX Supplement	Thermo Fisher Scientific	Cat. No. 35050038
Hank's Balanced Salt Solution, Ca^{2+}, Mg^{2+} (HBSS)	Thermo Fisher Scientific	Cat No. 24020091
Horse Serum, heat inactivated	Thermo Fisher Scientific	Cat No. 26050088
Kynurenic acid	Sigma-Aldrich	Cat. No. K3375
Millicell cell culture inserts	Merck Millipore	Cat. No. PICM03050
Millipore membrane confetti	Merck Millipore	Cat. No. FHLC04700
Minimum Essential Medium (MEM)	Thermo Fisher Scientific	Cat No. 11700077
Equipment	Company	
McIlwain tissue chopper	Ted Pella	California, USA

2.1.1. Slice cultures made from C57BL/6N wild-type mice

Mouse pups at postnatal day (P)3-7 were decapitated and the brain was removed and placed quickly into ice-cold preparation medium (97 mL Hank's balanced salt solution, 2.5 mL 20% glucose, and 1 mL 100 mM kynurenic acid, pH adjusted to 7.4). Both hippocampi were dissected with excess EC attached and transferred to a tissue chopper (McIlwain tissue chopper) platform where they were sectioned into 300 µm-thick slices perpendicular to the long-axis of the hippocampus. Slices were washed off the stage into ice-cold preparation medium. Slices were selected based on cell lamination and slice morphology and transferred onto sterile Millipore membrane confetti pieces that were placed on top of 6-well membrane inserts in pre-equilibrated culture medium (22.44 mL ddH2O, 25 mL 2xMEM, 25 mL BME, 1 mL GlutaMAX, 1.56 mL 40% Glucose, 25 mL horse serum). Residual medium around the slices was aspirated using a P200 pipette. A maximum of 12 hippocampal slices were cultured per animal and a maximum of four slices plated per membrane insert. Slices were maintained for 14 and 28 days at 37° C and 5% CO_2 with a medium change every 2-3 days.

2.1.2. Slice cultures made from Munc13-deficient and control mice

For the generation of Munc13-1 (Unc13A) and Munc13-2 (Unc13B) double knock-out (DKO) (Augustin et al., 1999; Varoqueaux et al., 2002) and control (CTRL) littermates mice with Unc13A[+/-] (Munc13-1) Unc13B[+/-] (Munc13-2) genotype were bred with Unc13A[+/-] Unc13B[-/-]. Only mice with the genotypes Unc13A[+/-] Unc13B[+/-] and Unc13A[+/+] Unc13B[+/-] were used for tomographic analysis of CTRL animals. Slice cultures from Munc13-1/2 DKO and CTRL littermates were prepared at embryonic day (E)18 due to the severe perinatal lethal phenotype of the Munc13-1/2 DKO (Varoqueaux et al., 2002). The slice culture procedure was identical to that of wild-type mice outlined in the previous section. However, due to the earlier developmental age of the pups, the hippocampi were smaller and therefore a maximum of 8 hippocampal slices were cultured per animal. Slices were maintained for 28 days at 37° C and 5% CO_2 with a medium change every 2-3 days.

2.2. High-pressure freezing, automated freeze substitution, and sample preparation for electron microscopy

2.2.1. High-pressure freezing of organotypic slice cultures

Table 4. Reagents and equipment for high-pressure freezing.

Reagents used for high-pressure freezing	Source	Catalog number
1-Hexadendene	Sigma Aldrich	Cat. No. 52276-5mL
Bovine Serum Albumin	Biomol	Cat. No. 01400.1
Bovine Serum Albumin	Sigma Aldrich	Cat. No. A2153
Equipment	Company	Source
High-pressure freezing device	Leica	Wetzlar, Germany

Slices were transferred to fresh culture medium 24 hours prior to fixation. Slices were transferred to pre-equilibrated slice culture medium and excess membrane was carefully removed with a razor blade. Slices were briefly submerged in slice culture medium containing 20% bovine serum albumin (BSA), which acts as a non-penetrating cryoprotectant, and loaded into the 100 µm-deep cavity of an aluminum planchette (type A, Leica Cat# 16770126, outer diameter 6 mm, inner cavity depth 100 µm). The planchette was then transferred to the middle plate on the HPF device (Leica HPM100 LS) and covered with the flat side of a type B aluminum planchette (Leica Cat# 16770127) coated with 1-Hexadecene (Sigma Aldrich), which facilitates separation of planchettes at post-cryofixation steps. Since gas is compressible, and the freezing process occurs at ~2000 bar atmospheric pressure, care was taken not to introduce air bubbles into the planchette cavity at any stage of the process. After HPF, cryofixed samples were stored in liquid nitrogen until further processing.

2.2.2. Acute brain slice preparation

Wild-type animals at P18 were anaesthetized, quickly decapitated, and brains were removed from the skull. Hippocampi were dissected from the cortices, placed on a tissue chopper, and 200 µm-thick sections were cut. Slices were removed from the tissue chopper and placed in HEPES-buffered artificial cerebrospinal fluid (ACSF) containing 20% BSA as a non-penetrating cryoprotectant. The CA3 and CA1 were isolated from the acute slice with a biopsy punch (diameter of 1.5 mm) and placed in the 200 µm-deep cavity of a 3 mm aluminum planchette (Leica Cat# 1677141 for type A). Hexadecene-coated lids (the flat side of type B 3 mm aluminum planchettes; Leica Cat# 1677142) were placed over the sample and quickly high-pressure frozen. The time between decapitation and HPF was no longer than 5 minutes.

2.2.3. Transcardial perfusion

Table 5. Reagents and equipment for chemical fixation.

Reagents for chemical fixation	Source	Catalog number
25% Glutaraldehyde	Electron microscopy sciences	Cat. No. 16220
Di-Sodium hydrogen phosphate dihydrate	Merck	Cat. No. 1.06580.1000
Paraformaldehyde	Serva	Cat. No. 31628.02
Sodium dihydrogen phosphate monohydrate	Merck	Cat. No. 1.06346.0500
Sodium cacodylate trihydrate	Sigma Aldrich	Cat. No. C0250-100G
Equipment	**Company**	
Leica Vibratome	Leica	Wetzlar, Germany

P28 wild-type mice were given an intraperitoneal injection of a lethal dose of Avertin (2,2,2,-Tribromoethanol). Once the mice were deeply anaesthetized, they were transcardially perfused first with 0.9% sodium chloride followed by one of two fixatives: Perfusion Fixative 1 (PF1): ice-cold 4% PFA, 2.5% GA in 0.1 M phosphate buffer (PB), pH 7.4 (Rollenhagen et al.,

2007); or Perfusion Fixative 2 (PF2): 37° C 2% PFA, 2.5% GA, 2 mM $CaCl_2$, in 0.1 M cacodylate buffer (Chicurel and Harris, 1992). The brains were dissected from the mice and post-fixed with their respective fixative overnight at 4° C with gentile agitation. The brains were washed thoroughly with ice-cold 0.1 M PB (pH 7.4) before 100 µm-thick sections were cut using a vibratome (Leica VT1200S; amplitude of 1.5 mm, cutting speed 0.1 mm/sec). Sections were briefly stored in 0.1 M PB before HPF. The CA3 and CA1 regions were excised from the sections using a biopsy punch (diameter of 1.5 mm) and high-pressure frozen in 3 mm aluminum planchettes. Tissue was frozen in a mixture of 20% BSA dissolved in 0.1 M PB.

2.2.4. Immersion fixation of hippocampal slice cultures

Hippocampal slice cultures at days *in vitro* (DIV)28 were quickly removed from the cell culture incubator and immersed in one of two fixatives: Immersion Fixative 1 (IF1): ice-cold 4% PFA, 2.5% GA in 0.1 M PB, pH 7.4 (Rollenhagen et al., 2007); or Immersion Fixative 2 (IF2): 37° C 2% PFA, 2.5% GA, 2 mM $CaCl_2$, in 0.1 M cacodylate buffer (Chicurel and Harris, 1992). Slices immersed in IF1 were incubated overnight at 4° C with gentile agitation. Slices were immersed in IF2 at an initial temperature of 37° C and were slowly cooled to room temperature for one hour with gentile agitation and then at 4°C overnight. The slices were then washed thoroughly with 0.1 M PB (pH 7.4). The CA3 region was isolated from the fixed slice with a biopsy punch (1.5 mm in diameter) and cryofixed in 20% BSA dissolved in 0.1 M PB (pH 7.4) as a non-penetrating cryoprotectant. Untreated hippocampal slices from the same cultures were cryo-fixed in tandem at DIV29 to serve as controls (processed as described above).

2.2.5. Acute pharmacological silencing experiments

Table 6. Reagents used for pharmacological treatment of organotypic slices.

Pharmacological agents	Source	Catalog number
(-)-Bicuculline methochloride	Tocris Bioscience	Cat. No. 0131
Biocytin hydrochloride	Sigma-Aldrich	Cat. No. B1758
D-AP5	Tocris Bioscience	Cat. No. 0106
DCG-IV: (2S,1'R,2'R,3'R)-2-(2,3-dicarboxycyclopropyl) glycine	Tocris Bioscience	Cat. No. 0975
Forskolin	Sigma Aldrich	Cat. No. F3917-25mg
NBQX disodium salt	Tocris Bioscience	Cat. No. 1044
Tetrodotoxin	Tocris Bioscience	Cat. No. 1078
Tetrodotoxin citrate	Tocris Bioscience	Cat. No. 1069

The protocol for the application of pharmacological agents to cultured slices was based on a previously published protocol for the application of drugs to organotypic slice cultures (Studer et al., 2014). Wild-type organotypic slices at DIV14 were placed onto new, sterile membrane inserts in a 6-well plate containing fresh, pre-equilibrated organotypic slice culture medium supplemented with one of two drug cocktails: (1) T/N/A, comprising 1 µM tetrodotoxin (TTX) to block sodium propagated action potentials, 2 µM 2,3-Dioxo-6-nitro-1,2,3,4-tetrahydrobenzo[*f*]quinoxaline-7-sulfonamide (NBQX) to block excitatory postsynaptic α-amino-3-hydroxy-5-methyl-4-isoxazolepropionic acid (AMPA) receptors, and 50 µM D-(-)-2-Amino-5-phosphonopentanoic acid (D-AP5) to block excitatory postsynaptic N-methyl-D-aspartate (NMDA) receptors; and (2) T/D, comprising 1 µM TTX and 2 µM DCG-IV, an mGluR2 receptor agonist that specifically reduced mossy fiber synaptic transmission (Kamiya et al., 1996). The vehicle control (VC) used as a negative control condition comprised medium alone. Then, 50 µL of medium containing either T/N/A, T/D, or VC were pipetted onto slices and incubated at 37° C and 5% CO_2 for 10 minutes. Slices were then prepared for HPF as described above, with the exception that the cryoprotectant-supplemented medium used prior to freezing also contained the respective drug cocktails at the indicated concentrations.

2.2.6. Pharmacological manipulation of presynaptic cAMP

Wild-type organotypic slices at DIV28 were transferred onto a new membrane insert in a six-well plate containing pre-equilibrated slice culture medium supplemented with one of two drug cocktails: (1) T/D, comprising 1 µM TTX, 2 µM DCG-IV and 0.08% dimethyl sulfoxide (DMSO); or (2) T/F, comprising 1 µM TTX, 0.2% additional ddH$_2$O, and 25 µM forskolin, an activator of AC1 that causes the enhancement of mossy fiber synaptic transmission (López-García et al., 1996; Villacres et al., 1998; Weisskopf et al., 1994). The VC used as a negative control condition comprised 1 µM TTX, 0.08% DMSO, and 0.2% additional distilled water. Slices were incubated for 15 minutes at 37° C and 5% CO$_2$ and prepared for HPF as described above, with the exception that the cryoprotectant supplemented medium used prior to freezing also contained the respective drug cocktails at the indicated concentrations.

2.2.7. Automated freeze substitution

Frozen slices were processed for AFS according to published protocols (Imig and Cooper, 2017; Rostaing et al., 2006). Vitrified slices were removed from liquid nitrogen storage and made accessible to freeze substitution media by separating type A and type B aluminum planchettes with custom-designed cryo-forceps. Samples were submersed in liquid nitrogen during this process to prevent the crystallization of water molecules in the tissue. Type A planchettes containing vitrified slices were then placed in AFS buckets in EM-grade acetone (Electron Microscopy Services, Cat# 10015) at -90° C. Samples were then incubated for 99 hours in 0.1% tannic acid (Sigma Aldrich, Cat# 403040-100G) dissolved in EM-grade acetone at -90°. The samples were then fixed with 2% OsO$_4$ in acetone starting at -90° C and slowly ramping the temperature up to 4° C at a rate of 5° C per hour until the temperature reached -20° C (16 hours) and then at a rate of 10° C per hour (2 hours). Residual OsO$_4$ was thoroughly washed from the samples with pre-cooled EM-grade acetone before the samples were brought to room temperature for Epon epoxy resin infiltration.

2.2.8. Plastic embedding

For epoxy resin embedding, samples were incubated in Eppendorf capsules in increasing concentrations of Epon (21.4 g glycidether, Serva; 14.4 g, dodecenylsuccinic acid anhydride, DDSA, Serva; 11.3 g methylnadic anhydride, MNA, Serva; 840 µL, tris(dimethylaminomethyl) phenol, DMP-30, Electron microscopy services) resin dissolved in EM-grade acetone: 50% Epon (4-6 hours); 90% Epon (overnight). Samples were then transferred to fresh Eppendorf capsules and incubated in three exchanges of 100% Epon over two days. For polymerization steps, carrier planchettes containing the osmified slices were placed sample-side up on a parafilm-covered glass slide. An Epon-filled gelatin capsule (size 00; Plano; Cat# G29218) containing a small specimen label was inverted over the sample and polymerized by baking at 60°C for 24-36 hours. Polymerized blocks were trimmed using a diamond-tipped milling device (Leica Reichert Jung Ultratrim) and planchettes were carefully removed with a razor blade to expose the tissue for subsequent ultramicrotomy.

2.2.9. Ultramicrotomy and contrasting

An Ultracut UCT ultramicrotome (Leica) equipped with diamond knives (Diatome, jumbo and ultra 45°) was used to acquire plastic-embedded tissue sections at three different thicknesses: 60 nm ultrathin sections were collected on for 2D ultrastructural analyses; 200 nm sections for 3D electron tomography; and 500 nm sections for low-magnification orientation using light microscopy. For ultrastructural analyses, sections were collected on formvar-coated grids (Electron Microscopy Services; 100 square copper; Cat# G2100C) and stored in grid boxes until further use. For 2D ultrastructural analysis, lipid bilayer contrast was enhanced by floating 60 nm-thick grid-mounted sections on solutions of 1% aqueous uranyl acetate for 30 minutes followed by 0.3% Reynold's lead citrate for 2 minutes. For 3D electron tomography, gold fiduciary markers were deposited on both surfaces of 200 nm-thick grid-mounted sections with 10 nm gold-conjugated Protein A (Cell Microscopy Center, Utrecht, The Netherlands). To obtain light microscopic overviews of sectioned tissue, 500 nm-thick sections were dried on glass slides, and then contrasted with methylene blue Nissl stain to visualize patterns of cell body lamination.

2.3. Electron microscopy

Table 7. Reagents and equipment for sample processing and preparation for electron

Reagents	Source	Catalog number
2,4,6-Tris(dimethylaminomethyl)phenol (DMP-30)	Electron microscopy sciences	Cat. No. 13600
2-Dodecenylsuccinic acid anhydride (DDSA)	Serva	Cat. No. 20755.02
Acetone	Electron microscopy sciences	Cat. No. 10015
Glycidether 100	Serva	Cat. No. 21045.02
Lead (II) Nitrate	Merck	Cat. No. 1.07398.0100
Methylnadic anhydride (MNA)	Serva	Cat. No. 29452.02
Osmium tetroxide	Electron microscopy sciences	Cat. No. 19132
Protein A (ProtA) coupled to 10 nm gold particles	Cell Microscopy Core Products, University Medical Center Utrecht, The Netherlands	
Sodium Citrate	Calbiochem	Cat. No. 567446
Tannic Acid 0.1%	Sigma Aldrich	Cat. No. 403040-100G
Uranyl Acetate	SPI Supplies	Cat. No. 2624
Equipment	Company	Source
Leica Vibratome	Leica	Wetzlar, Germany
EM AFS2	Leica	Wetzlar, Germany
Leica Reichert Jung Ultratrim	Leica	Wetzlar, Germany

2.3.1. Transmission electron microscopy imaging and analysis

For 2D ultrastructural analyses, images were acquired on an 80 kV Leo912 TEM (Zeiss) equipped with a sharp:eye CCD camera (Tröndle, TRS) and iTEM (Olympus Soft Imaging Solutions GmbH) software. Schaffer collateral and mossy fiber synapses were identified according to their distinct morphologies in montaged images acquired at 5,000x magnification (image pixel size = 2.269 nm) from CA1 *stratum radiatum* and CA3 *stratum lucidum*, respectively. The CA1, CA3b and CA3c regions containing Schaffer collateral and mossy fiber synapses, respectively, were acquired at 20,000x magnification (image pixel size = 0.592 nm). Material exhibiting signs of freezing damage, i.e. ice crystal formation, were identified according to published qualitative criteria (Möbius et al., 2010) and excluded from further analysis.

In forskolin-treated and corresponding VC slices (see *Pharmacological manipulation of presynaptic cAMP*), the following morphological parameters were quantified from 2D electron micrographs using the IMOD package (Kremer et al., 1996) in combination with the *imodinfo* and *mtk* programs: (i) presynaptic bouton area and membrane circumference, (ii) active zone number and length, (iii) spine area and presynaptic contact length, and (iv) mitochondrial number, area, and circumference.

2.3.2. Electron tomography and data analysis

Table 8. Imaging software and equipment for 2D electron microscopy and 3D electron tomography.

Imaging and Analysis Software	Supplier	Source
IMOD software	Kremer et al., 1996	
iTEM software	Emsis GMBH	Emsis GMBH
SerialEM software	University of Colorado, Boulder, Colorado, US	http://bio3d. colorado.edu/SerialEM/
Equipment	Company	Source
Leo912 Transmission electron microscope	Zeiss	Jena, Germany
JEM 2100 transmission electron microscope	Jeol	Tokyo, Japan

Electron tomograms from Schaffer collateral and mossy fiber active zone release sites were generated on a 200 kV JEM2100 TEM (JEOL) equipped with an Orius SC1000 digital camera (Gatan). Single-axis tilt series (-60° to +60°, 1° increments) were acquired at 30,000x magnification with a 2x binning factor (image pixel size = 0.52 nm) using Serial EM software (Mastronarde, 2005). Only synapses in which the synaptic cleft was clearly visible at 0° tilt were selected for reconstruction using the weighted back-projection feature of the IMOD package (Kremer et al., 1996) and a 3x binning factor (tomogram voxel dimensions x,y,z = 1.554 nm). The location and dimensions of reconstructed synaptic active zones were identified according to four morphological criteria: 1) an accumulation of presynaptic vesicles, 2) a directly apposing postsynaptic density, 3) a widening of the synaptic cleft, and 4) the presence of electron dense trans-synaptic cleft material (Gray, 1959; High et al., 2015; Palay, 1956). These criteria were necessary in some cases for analysis of mossy fiber synapses due to the presence of multiple active zones in proximity to one another and that additional protein contrasts were not used for sections imaged with electron tomography. Vesicles within 100 nm of the active zone were segmented manually as size-matched spheres positioned at the vesicular midline, i.e. the tomographic slice of largest vesicular diameter.

The diameter of segmented spheres was adjusted to correspond to the outer leaflet of the vesicle lipid bilayer. Non-spherical organelles (e.g. endoplasmic reticulum, tubular endosomal intermediates) were occasionally observed in tomographic reconstructions, but excluded from the analysis. Active zones were segmented as open contours corresponding to the inner leaflet of the presynaptic plasma membrane.

Vesicle radii and active zone surface areas were extracted from segmented tomograms using the *imodinfo* program of the IMOD package (Kremer et al., 1996). The closest approach of vesicles to the active zone was measured in Euclidean space using the *mtk* program of the IMOD package (Kremer et al., 1996). Docked synaptic vesicles in direct contact with the active zone membrane were manually quantified according to the criterion that no measurable distance was observed between the outer leaflet of the vesicle lipid bilayer and the inner leaflet of the presynaptic membrane (i.e. when the dark pixels corresponding to the vesicular membrane were contiguous with those of the plasma membrane). The number of vesicles measured within discrete distances from the active zone membrane [i.e. 0-2 (docked), 0-40, and 0-100] were normalized to the active zone area and reported as a spatial density (i.e. number of vesicles per $0.01\,\mu m^2$ active zone). Vesicles were classified into three morphological categories according to their diameter and lumenal content: clear-cored vesicles with a diameter less than 60 nm were classified as synaptic vesicles; clear-cored vesicles with a diameter exceeding 60 nm were classified as giant vesicles; and vesicles with a prominent electron-dense core were classified as DCVs irrespective of their diameter.

2.4. RRP calculations

In mossy fiber-CA3 spine synapses, my calculations of mean docked vesicle numbers per active zone and per mossy fiber bouton were based on the mean number of docked vesicles per unit of active zone area (0.97 synaptic vesicles, 0.1 giant vesicles and 0.05 DCVs per $0.01\,\mu m^2$ active zone area) quantified in this work, as well as previously published estimates of the mean active zone surface area ($0.12\,\mu m^2$) and mean active zone number (29.75 active zones) per bouton in P28 rat mossy fiber boutons (Rollenhagen et al., 2007). I calculated the mean docked vesicle numbers per active zone (11.6 synaptic vesicles, 1.1 giant vesicles, 0.6 DCV) and per bouton (~345 synaptic vesicles, 33 giant vesicles, 18 DCVs) in mossy fiber synapses from DIV28 slice cultures as well as the mean number of total membrane-proximal (within 0-40 nm of the active zone; 25.1 synaptic vesicles, 3 giant vesicles, 2.3 DCVs) vesicles per active zone.

I extrapolated the surface area and volume of all docked vesicles according to their size and morphological classification (mean diameters: synaptic vesicles, 45.17 nm; giant vesicles, 85.77 nm; DCVs 74.41 nm). Based on a specific membrane capacitance of $1\,\mu F/cm^2$ (Hallermann et al., 2003), I estimated that, (i) the fusion of all docked vesicles, irrespective of their type, would correspond to a membrane capacitance increase of ~33 fF per mossy fiber bouton, and (ii) that the fusion of all vesicles within 0-40 nm of the active zone membrane would correspond to a membrane capacitance increase of ~80 fF per mossy fiber bouton. A limitation of this approach is that my data, which is based exclusively on tomograms from mossy fiber-CA3 spine synapses, does not take into account synapses made via filopodial extensions (Acsády et al., 1998).

2.5. Electrophysiology

Table 9. Software used for electrophysiology acquisition and analysis.

Software	Supplier		
GraphPad Prism 5 and 7	GraphPad Software		
Axograph X 1.3.3	John Clements		
Patchmaster v2X80	HEKA/Harvard Bioscience		

2.5.1. Miniature excitatory postsynaptic currents in CA3 pyramidal neurons in slice culture at DIV14

Performed by Dr. Bekir Altas

All recordings of miniature EPSCs (mEPSCs) from CA3 pyramidal cells were performed in wild-type organotypic slice cultures at DIV14. Prior to recording, slices were incubated for 30 minutes in an interface chamber with carbogen-saturated ACSF (120 mM NaCl, 26 mM NaHCO₃, 10 mM D-glucose, 2 mM KCl, 2 mM MgCl₂, and 2 mM CaCl₂, and 1 mM KH₂PO₄ with an osmolarity of 304 mOsm). One or two CA3 pyramidal cells were then whole-cell voltage clamped using a glass pipette (2.5-3.0 MΩ) filled with internal solution (100 mM KCl, 50 mM K-gluconate, 10 mM HEPES, 4 mM ATP-Mg, 0.3 mM GTP-Na, 0.1 mM EGTA, and 0.4% biocytin, pH 7.4 with an osmolarity of 300 mOsm) and the holding potential was set at -70 mV using an EPC-10 amplifier [Patchmaster 2 software (HEKA/Harvard Bioscience)]. For measurements of mEPSC amplitudes and frequencies, slices were initially perfused with 1 µM TTX and 10 µM bicuculline and mEPSCs were then recorded for 10 minutes, after which the slices were perfused with 1 µM TTX, 10 µM bicuculline, and 2 µM DCG-IV for 15 minutes to record DCG-IV insensitive mEPSCs. Measurements of all mEPSCs (TTX/bicuculline) were recorded in two-5-minute epochs, while measurements of non-mossy fiber mEPSCs (TTX/bicuculline/DCG-IV) were recorded in three-5 minunte epochs. The last epochs of each recording were used for mEPSC analysis. All electrophysiological traces were analyzed using Axograph X software (AxoGraph Scientific) using a template fit algorithm for automatic event detection (Jonas et al., 1993; Pernía-Andrade et al., 2012). After recordings, slices were fixed and biocytin-filled CA3 pyramidal cell were stained with Alexa Fluoro-555-labeled streptavidin (see Light Microscopic Analysis section for detailed procedure).

Only cells exhibiting a reduction in the mEPSC frequency upon application of DCG-IV were analyzed (reduction range 16.9–88.6%; mean 59.2%). The threshold for mEPSC detection was set to 8 pA. The amplitude distribution of mEPSC events was plotted (1 pA bins) for epochs prior to and following the application of DCG-IV. The remaining DCG-IV-insensitive events were subtracted from the events recorded prior to drug application in the respective bins to isolate the DCG-IV-sensitive component (likely of mossy fiber origin). Operating on the assumption that DCG-IV-sensitive mEPSC events result from the fusion of docked vesicles at mossy fiber active zone release sites, the statistical mode of mEPSC amplitudes (10 pA) was correlated with the statistical mode of docked vesicle diameters (44 nm). Since vesicle diameters were measured between the outer leaflets of the vesicle lipid bilayer, lumenal volumes were calculated by subtracting the thickness of the lipid bilayers (each approximately 4 nm-thick as measured from center-to-center of inner and outer leaflets). Based on a previous, conceptually analogous study (Bruns et al., 2000), I assumed uniform neurotransmitter loading irrespective of vesicle size and negligible saturation of postsynaptic receptors to estimate that an mEPSC amplitude of approximately 30 pA would arise from the fusion of a vesicle with a diameter of 60 nm (size threshold for classification of a giant vesicle).

2.5.2. Release probability and short-term plasticity of Schaffer collateral and mossy fiber synapses in slice cultures at DIV14 and DIV28

Performed by Dr. Chungku Lee

Whole-cell voltage-clamp recordings of evoked EPSCs in CA1 and CA3 neurons in wild-type organotypic slice cultures at DIV14 and DIV28 were performed by extracellular stimulation of Schaffer collaterals and mossy fibers, respectively. Short current pulses (0.1 ms, 100-1000 µA) were applied using an ACSF-filled bipolar theta glass electrode (Science Products, Hofheim, Germany). Whole-cell recordings were made at -70 mV in the presence of 100 µM picrotoxin to block inhibitory currents. Borosilicate glass pipettes (4-5 MΩ; P.Clamp Glass #0010, WPI) were filled with intracellular solution (115 mM Cs-methanesulfonate, 10 mM HEPES, 10 mM EGTA, 5 mM MgCl2, 5 mM QX-314, 4 mM Na2-ATP, 0.3 mM Na2-GTP, and 0.2 % biocytin, pH 7.4). The serial resistance was compensated by 45%–80% and only cells with serial resistances below 12 MΩ were analyzed. All experiments were recorded by an EPC-10 USB double amplifier and Patch-Master (Ver. 2X73.5) software (HEKA electronics) and the obtained data were analyzed using Axograph (Ver.1.5.4) software (Axograph Scientific). Statistical analysis of data was performed with GraphPad Prism (versions 5 or 7) software (GraphPad Software Inc., La Jolla, CA).

2.6. Light microscopic analysis

Performed with help from Dr. Benjamin Cooper and Manuela Schwark

2.6.1. Sample preparation for confocal microscopy

To demonstrate the correct anatomical organization of the mossy fiber pathway in organotypic slices, cultures were immersion-fixed in 4% PFA in 0.1 M PB (pH 7.4) overnight at 4° C. Slices were washed in 0.1 M PB (pH 7.4) and then permeabilized and blocked in 10% normal goat serum (NGS), 0.3% Triton X-100, and 0.1% cold water fish skin gelatin in 0.1 M PB (pH 7.4) overnight at 4° C. Whole-mount immunolabelling was performed by incubating slices overnight at 4° C in 5% NGS, 0.3% Triton X-100 and 0.1% fish skin gelatin in 0.1 M PB (pH 7.4) containing primary antibodies against synaptic vesicle clusters within the synaptic terminals of mossy fiber projections [polyclonal rabbit anti-synaptoporin, Synaptic Systems (Cat# 102

003), 1:1000 dilution] and cell bodies and dendritic arborizations [polyclonal chicken anti-MAP2, Novus Biologicals (Cat# NB300-213), 1:600 dilution]. Slices

Table 10. Reagents and equipment used for immunostaining and light microscopy.

Antibodies	Dilution	Source	Catalog number
Chicken anti-MAP2 antibody	1:600	Novus	Cat. No. NB 300-213
Goat anti-Chicken IgG secondary antibody, Alexa 488	1:1000	Thermo Fisher Scientific	Cat. No. A-11039
Goat anti-Rabbit IgG secondary antibody, Alexa 555	1:1000	Thermo Fisher Scientific	Cat. No. A21429
Goat anti-Rabbit IgG secondary antibody, Alexa Fluor 488	1:1000	Thermo Fisher Scientific	Cat. No. A-11008;
Mouse anti-bassoon antibody	1:400	Enzo	Cat. No. ADIVAM-PS003-F
Rabbit anti-synaptoporin antibody	1:500	SynapticSystems	Cat. No. 102 003
Goat anti-Mouse secondary antibody, ATTO647N	1:100	Rockland	Cat. No. 610-156-12
Goat anti-Rabbit secondary antibody, START580	1:100	Abberior	Cat. No. ST580-1002-500UG
Alexa Fluor 555 streptavidin conjugate	1:500	Thermo Fisher Scientific	Cat. No. S32355; RRID: AB_2571525
DAPI: 4',6-Diamidine-2'-phenylindole dihydrochloride	1:1000	Sigma-Aldrich/Roche	Cat. No. 10236276001
Imaging and Analysis Software	**Supplier**	**Source**	
ImageJ	National Institutes of Health	https://imagej.nih.gov/ij	
Leica LAS AF	Leica Microsystems	http://www.leica-microsystems.com	
Equipment	**Company**	**Source**	
SP5 Confocal Microscope	Leica	Wetzlar, Germany	
Zeiss Apotome image Z.1	Zeiss	Jena, Germany	

were washed in 0.1 M PB (pH 7.4) and primary antibodies were visualized by a two-hour incubation at room temperature in 5% NGS, 0.1% Triton X-100 and 0.1% fish skin gelatin in 0.1 M PB (pH 7.4) containing goat anti-rabbit Alexa 555 [Thermo Fisher (Cat# A21429), dilution 1:1000] and goat anti-chicken Alexa 488 [Thermo Fisher (Cat# A-11039), dilution

1:1000]. Following final washing steps in 0.1 M PB (pH 7.4), slices were floated onto Superfrost™ glass slides with the membrane confetti in contact with the slide and Menzel-Gläser #1.5 glass coverslips were mounted using Aqua-Poly/Mount mounting medium (Polysciences, Inc., Cat# 18606-20).

To resolve active zone release sites within mossy fiber boutons, slices were removed from culture inserts and fixed by overnight immersion in 4% PFA in 0.1 M PB (pH 7.4). Slices were washed in 0.1 M PB (pH 7.4) and then cryoprotected in an increasing sucrose gradient (from 10% to 30%) in 0.1 M PB (pH 7.4) until saturation and then positioned slice-side down (confetti-side up) as flat as possible on the inner base of a quadratic 10 x 10 x 10 mm form made out of aluminum foil. The form was carefully filled with liquid Tissue-Tek® OCT compound (Sakura, Cat# 4583) and then rapidly frozen on a liquid nitrogen-cooled aluminum block. Frozen slices were stored at -80° C until cryosectioning. Prior to cryosectioning, blocks were pre-equilibrated in the cryostat at -20° C the night before sectioning. The aluminum foil was removed and the frozen OCT block was mounted slice-side up on a specimen stub with OCT in a precooled (specimen holder, -18° C; chamber, -18° C) cryostat (Leica CM3050 S). Once the temperature of the embedded slice had equilibrated, unnecessary OCT compound was trimmed away with a razor blade and 10 μm-thick cryosections were made through the organotypic slice and thaw-mounted on Superfrost™ slides. Slides were air-dried at room temperature for 30 minutes and a hydrophobic pen (DAKO, Cat# S2002) was used to delineate the border of the slide surface. The hydrophobic pen helped minimize the volume of antibodies used for immunostaining. Slides were washed briefly in 0.1 M PB (pH 7.4) and incubated for 90 minutes at room temperature in 10% NGS, 0.3% Triton X-100, and 0.1% fish skin gelatin in 0.1 M PB (pH 7.4). Slices were then incubated overnight at 4° C in 3% NGS, 0.3% Triton X-100 and 0.1% fish skin gelatin in 0.1 M PB (pH 7.4) containing primary antibodies for the detection of synaptic vesicle clusters within the synaptic terminals of mossy fiber projections [polyclonal rabbit anti-synaptoporin, Synaptic Systems (Cat# 102 003), 1:1000 dilution] and presynaptic active zones [monoclonal mouse anti-bassoon, Enzo Life Sciences (Cat# SAP7F407), 1:400 dilution]. Slices were washed in 0.1 M PB (pH 7.4) and primary antibodies were visualized by two-hour incubation at room temperature in 5% NGS, 0.1% Triton X-100 and 0.1% fish skin gelatin in 0.1 M PB (pH 7.4) containing goat anti-rabbit Alexa 488 [Thermo Fisher (Cat# A11008), dilution 1:1000] and goat anti-mouse Alexa 555 [Thermo Fisher (Cat# A21424), dilution 1:1000]. Following a brief wash in 0.1 M PB, slides were dipped in distilled water and Menzel-Gläser #1,5 coverslips were mounted using Aqua-Poly/Mount mounting medium (Polysciences, Inc., Cat# 18606-20).

Confocal light microscopic analysis of biocytin-filled CA3 pyramidal cells was performed to validate that electrophysiological recordings were of the correct filled cells, as assessed by the anatomical location within the hippocampal subfields and morphological features (i.e, pyramidal soma, presence of large, complex spines in the proximal regions of apical dendritic arborizations) (see Figure 5 D). Immediately following mEPSC recordings and removal of the patch pipette, biocytin-filled CA3 pyramidal cells (see Electrophysiology section above for detailed procedure) were fixed for light microscopic analysis by overnight immersion of the slice in 4% PFA in 0.1 M PB (pH 7.4). Slices were washed in 0.1 M PB (pH 7.4) and then incubated overnight at 4° C in 10% NGS, 0.3% Triton X-100, and 0.1% fish skin gelatin in 0.1 M PB (pH 7.4). Biocytin-filled cells were visualized by incubation of slices for three hours at room temperature in streptavidin-Alexa 555 [1:500 dilution] in 5% NGS, 0.1% Triton X-100 and 0.1% fish skin gelatin in 0.1 M PB (pH 7.4). Slices were washed in 0.1 M PB (pH 7.4) and cell nuclei were stained by a 30-minute incubation in DAPI [300 nM in 0.1 M PB]. Following final washing steps in 0.1 M PB (pH 7.4), slices were floated onto Superfrost glass slides with the membrane confetti in contact with the slide and Menzel-Gläser #1,5 glass coverslips were mounted using Aqua-Poly/Mount mounting medium (Polysciences, Inc., Cat# 18606-20).

2.6.2. Confocal imaging

Confocal laser scanning micrographs were acquired with a Leica TCS-SP5 confocal microscope equipped with a tunable white light laser, a resonant scanner, hybrid GaAsP detectors, and a motorized stage. Tiled z-series were acquired with (i) a HCX PL APO 40.0x (NA=1.25) oil immersion objective to generate low magnification overviews of entire organotypic slices (pinhole = 3.0 AU, voxel size x, y, z = 0.3, 0.3, 2 µm (Figure 5 A) and reconstructions of biocytin-filled pyramidal neurons within CA3 *stratum pyramidale* (pinhole = 1.0 AU, voxel size x, y, z = 95, 95, 335 nm) (Figure 5 D), or with (ii) a HCX PL APO CS 100x (NA=1.4) oil immersion objective to visualize mossy fiber terminals within CA3 *stratum lucidum* (pinhole = 1.0 AU, voxel size x, y, z = 89, 89, 130 nm) and high magnification reconstructions of complex postsynaptic spines (thorny excrescences) emerging from the proximal dendrites of biocytin-filled CA3 pyramidal neurons (pinhole = 0.5 AU, voxel size x, y, z = 47, 47, 130 nm). For illustration purposes thorny excrescences were subjected to spatial deconvolution by use of two ImageJ (National Institutes of Health; Bethesda, MD) plugins: point spread functions were generated using Diffraction point spread function 3D plugin and iterative deconvolution was performed with the Richardson-Lucy algorithm (DeconvolutionLab plugin; Biomedical Imaging Group, EPFL; Lausanne, Switzerland).

2.6.3. Stimulated emission depletion microscopy

Imaging performed by Sinem Sertel

Wild-type hippocampal slice cultures at DIV28 were treated for 15 minutes with 1 µM TTX to block spontaneous network activity as a VC, or TTX with 25 µM forskolin to activate AC1 and increase presynaptic cAMP (Huang et al., 1994a; Villacres et al., 1998). Slices were transferred to a new Millipore membrane insert containing the respective drug mixtures and 50 µl of drug-containing medium was pipetted onto each slice. The plate was then placed in the incubator for 15 minutes, after which the slices were removed from the inserts and were immersed in 4% PFA in 0.1 M PB (pH 7.4) for one hour at 4° C. Slices were prepared for cryosectioning as described above with some exceptions. Sections (10 µm-thick) were thaw-mounted on Superfrost™ slides, blocked and permeabilized for one hour at room temperature in 5% NGS, 0.3% Triton X-100 and 0.1% fish skin gelatin in 0.1 M PB (pH 7.4), and were then incubated overnight at 4° C in 3% NGS, 0.1% Triton X-100, and 0.1% fish skin gelatin in 0.1 M PB (pH 7.4) containing primary antibodies for synaptoporin [polyclonal rabbit anti-synaptoporin, Synaptic Systems (Cat# 102 003), 1:500 dilution] and bassoon [monoclonal mouse anti-bassoon, Enzo Life Sciences (Cat# SAP7F407), 1:400 dilution]. Sections were washed in 0.1 M PB (pH 7.4) to remove primary antibodies and were then incubated for two hours at room temperature in 3% NGS, 0.1% Triton X-100 and 0.1% FSG in 0.1 M PB (pH 7.4) containing goat anti-mouse ATTO647N [Rockland (Cat# 610-156-12), dilution 1:100] and goat anti-rabbit STAR580 [Abberior (Cat# ST580-1002-500UG), dilution 1:100]. Following a brief wash in 0.1 M PB, slides were dipped in distilled water and Menzel-Gläser #1,5 coverslips were mounted onto them, covering the sections, using Mowiol mounting medium (Merk Millipore; Cat# 475904).

Dual stimulated emission depletion (STED) imaging was performed on bassoon and synaptoporin immunelabelling [using an Expert Line STED (Abberior) instrument based on an IX83 inverted microscope (Olympus)]. The images were analyzed with a Matlab script

(Mathworks) written by Sinem M. Sertel. The synaptoporin and bassoon images were thresholded to remove background and to enable the recognition of the different objects (spots). The positions of synaptoporin and bassoon objects were then calculated, and for each synaptoporin object we determined the overlap with the bassoon objects found near it (within 120 nm). The total area of bassoon objects per synaptoporin object was then determined, and the mean object value per image was plotted using GraphPad Prism 7.

2.7. Quantification and statistical analysis

Data are represented as mean ± standard error of the mean (SEM) unless indicated otherwise. Statistical analyses were carried out using GraphPad Prism software 7 (* when p<0.05; ** when p<0.01, and *** when p<0.001). For comparisons of two conditions (i.e. Schaffer collateral and mossy fiber synaptic profiles from DIV14 wild-type slice cultures; Figure 9 G-M) statistical differences were determined by an unpaired t-test when the data set was normally distributed as determined by a Kolmogorov-Smirnov normality test, and by a Mann-Whitney unpaired t-test if the data were not normally distributed. For pharmacological manipulation experiments, statistical significance was tested by one-way analysis of variance with Bonferroni correction as a post-test if the data set was normally distributed. If the data set was not normally distributed, Kruskal-Wallis analysis of variance test with a Dunn's comparison of all columns was performed to assess statistical significance. For the electron tomography experiments, the number of active zones analyzed for each experiment (n), the number of slice cultures or animals used (N), and all EM data are summarized in Tables 15-28 and 30-31. Statistics were performed based on the number of active zones for each sample with the exception of docked vesicle diameters and unattached giant vesicles and DCVs. In the latter scenarios, the number of vesicles was used for statistical analysis and is noted in parentheses. For the 2D EM analysis of mossy fiber synaptic profiles, the number of boutons (n), the number of slice cultures or animals used (N), and all data are summarized in Tables 33-34. Electrophysiological analyses of mEPSC recordings were performed on 28 cells from two independent wild-type slice cultures at DIV14. Statistical difference for electrophysiological experiments measuring mEPSCs was determined by Wilcoxon matched pairs signed rank tests. Electrophysiology of Schaffer collateral and mossy fiber evoked synaptic transmission was performed on three independent slice cultures. Evoked Schaffer collateral synaptic transmission was recorded from 12 cells at DIV14 and 9 cells at DIV28.

Evoked mossy fiber synaptic transmission was recorded from 11 cells at DIV14 and 9 cells at DIV28. For the short-term synaptic plasticity electrophysiology experiments, statistical comparisons between two groups of data were made using two-tailed unpaired Student's t-test. P-values less than 0.05 were considered significant for single and multiple comparisons, respectively.

3. Results

3.1. Differences in Schaffer collateral and mossy fiber synapse release probability are paralleled by synapse-specific differences in synaptic vesicle organization

My primary objective was to perform a comparative 3D ultrastructural analysis of Schaffer collateral and mossy fiber synapses at rest to test the hypothesis that differences in synaptic vesicle organization at active zones contribute to their different neurotransmitter release properties, i.e. release probability and short-term plasticity. To verify that the hippocampal organotypic slices used in my experiments serve as an appropriate model system to address this aim, I performed experiments to investigate, (i) whether the anatomical development and target specificity of the mossy fiber projection remains intact during the slice culture period, and (ii) whether the functional neurotransmitter release properties of Schaffer collateral and mossy fiber synapses described in acute brain slices are also preserved in cultured slices. Having established the anatomical and functional integrity of the organotypic slice culture system, I then compared the 3D spatial organization of synaptic vesicles at active zone release sites in Schaffer collateral and mossy fiber synapses within the same slice to test the hypothesis that initial release probability at these synapses is codetermined by the availability of docked and primed vesicles.

3.1.1. Intact target specificity of the mossy fiber pathway in hippocampal organotypic slice cultures

To assess whether the target specificity of the mossy fiber projection from the granule cell layer to the *stratum lucidum* of the CA3 was preserved in our culture system, I performed whole-mount immunolabelling experiments using an anti-synaptoporin antibody to detect the synaptic vesicle channel protein synaptoporin, which is highly enriched in hippocampal granule cells (Grabs et al., 1994), and an antibody directed against microtubule-associated protein 2 (MAP-2) to visualize the cell bodies and dendrites of CA3 pyramidal cells (Figure 5 A, B) (De Camilli et al., 1984). Light microscopic analysis revealed a pattern of immunolabelling that is highly comparable to the *in vivo* organization of the hippocampus, in which synaptoporin puncta are restricted to the hilus, where mossy fiber collaterals form synaptic

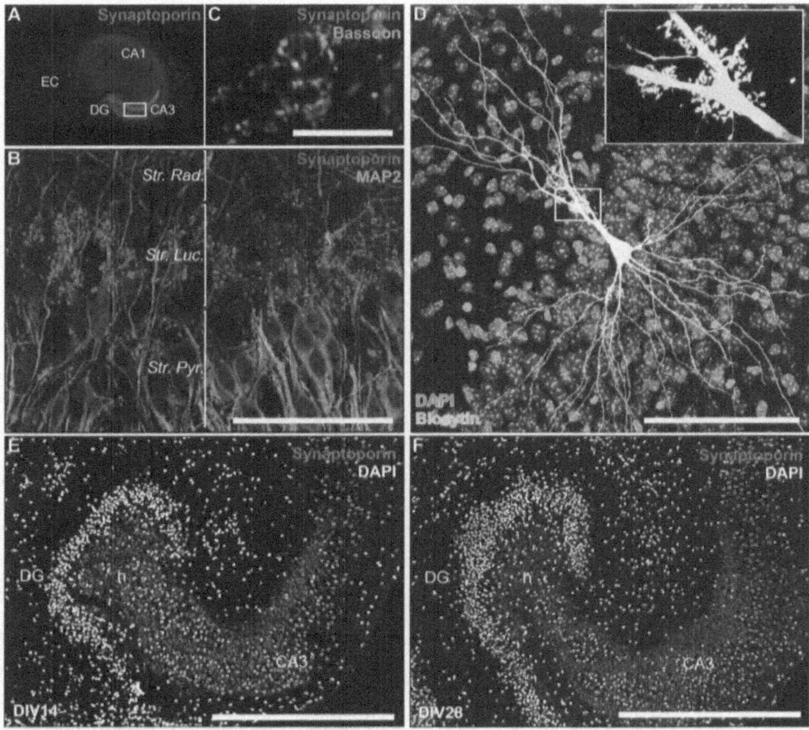

Figure 5. Light microscopic analysis of the mossy fiber-CA3 pathway in hippocampal organotypic slice cultures at DIV14 and 28.

(A) Widefield light microscopic overview of a wild-type hippocampal slice culture labelled against synaptoporin, a synaptic vesicle channel protein enriched in mossy fiber synapses. (B) Maximum projection of a confocal z-stack in which synaptoporin-labeled mossy fiber boutons (magenta) are seen to cluster around MAP-2 positive proximal dendrites (green) of pyramidal cells in CA3 *stratum lucidum*. (C) Single-plane confocal micrograph illustrating the high density of bassoon-labelled active zones (green) within large synaptoporin-positive mossy fiber boutons (magenta). (D) Maximum projection of a confocal z-stack through a CA3 pyramidal neuron filled with biocytin and visualized with streptavidin-conjugated Alexa 555 ("orange hot" LUT). Proximal dendrites are decorated with complex, multi-compartmental spines (thorny excrescences; insert) which are postsynaptic to mossy fiber-CA3 inputs. (E, F) Maximum projections of confocal z-stacks through organotypic slice cultures at DIV14 (E) and DIV28 (F). Synaptoporin (magenta) and DAPI (white) labelling is highly comparable at both developmental time-points, indicating that the expected target specificity of the mossy fiber-CA3 pathway remains intact throughout the analyzed culture period. No evidence of aberrant mossy fiber sprouting was observed. Abbreviations: EC, entorhinal cortex; DG, dentate gyrus; CA3, *cornu ammonis* area 3; CA1, *cornu ammonis* area 1; *Str.*, *stratum*; *Rad.*, *radiatus*; *Luc.*, *lucidum*; *Pyr.*, *pyramidale*; h, hilus; DIV, days *in vitro*. Scale bars: 5 μm, C; 100 μm, B and D; 500 μm, E and F.

contacts with hilar mossy cells, and to the *stratum lucidum* of the CA3, where mossy fiber boutons establish synaptic contacts with pyramidal cells. Within the CA3 *stratum lucidum*, large synaptoporin-positive puncta were observed to be clustered apposed to the primary dendrites of MAP-2-positive CA3 pyramidal neurons (Figure 5 B). To observe the distribution of active zones within mossy fiber terminals at a higher resolution, I performed a confocal light microscopic analysis of thin cryosections from cultured slices co-labelled with antibodies detecting synaptoporin and the active zone protein bassoon. Bassoon-immunoreactive puncta clustered at a high density within synaptoporin-positive mossy fiber boutons in the *stratum lucidum* (Figure 5 C). This finding indicates that mossy fiber boutons in slice cultures contain multiple active zones, which is a characteristic feature of this synapse type (Amaral and Dent, 1981; Chicurel and Harris, 1992; Rollenhagen et al., 2007). To test whether the postsynaptic organization of the mossy fiber-CA3 pyramidal cell connection is preserved in hippocampal slice cultures, biocytin-filled CA3 pyramidal neurons generated during patch-clamp recordings (performed by Dr. Bekir Altas) were post-labeled using streptavidin-conjugated Alexa-555. Consistent with descriptions of CA3 pyramidal neurons *in vivo* (Amaral et al., 2007; Chicurel and Harris, 1992; Frotscher et al., 2014; Gonzales et al., 2001), confocal light microscopic analysis revealed the presence of complex, multi-compartmental dendritic spines ("thorny excrescences") engulfing the primary dendrites of CA3 pyramidal neurons in cultured slices (Figure 5 D).

The pattern of synaptoporin immunoreactivity was highly comparable in thin cryosections from hippocampal slice cultures at DIV14 (Figure 5 E) and DIV28 (Figure 5 F), indicating that the anatomical organization of mossy fiber projection remains intact at both developmental time-points used for subsequent ultrastructural analyses. Synaptoporin-positive puncta were restricted to the hilus and *stratum lucidum* and no evidence of aberrant mossy fiber sprouting (Coltman et al., 1995) into the dentate granule cell layer was observed.

On the ultrastructural level (Figure 6), transmission electron micrographs of organotypic hippocampal slices exhibited an excellent preservation of synaptic ultrastructure as assessed by published criteria (Möbius et al., 2010). Consistent with previous reports in transcardially perfused rat brains (Chicurel and Harris, 1992; Rollenhagen et al., 2007), large mossy fiber boutons in the stratum lucidum formed three different types of contacts with postsynaptic CA3 pyramidal neurons, namely axo-spinous synaptic contacts onto thorny excrescences (Figure 6 B), axo-dendritic synaptic contacts onto the shaft of primary dendrites (Figure 6 C),

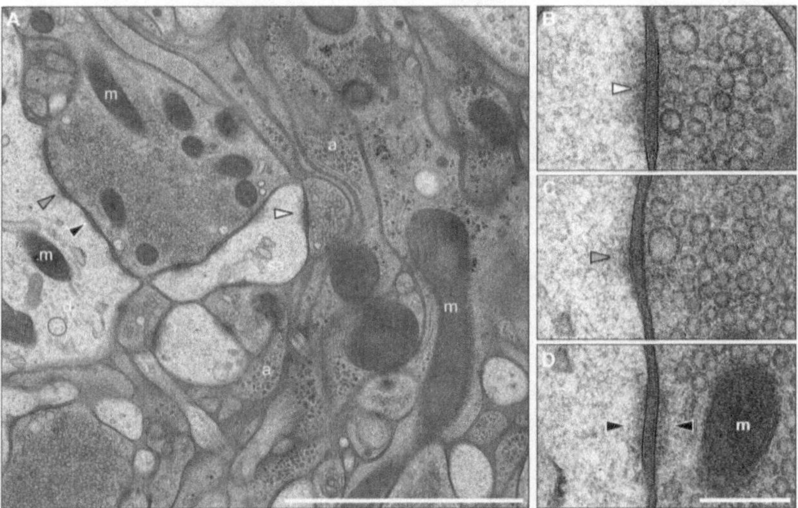

Figure 6. Ultrastructural organization of mossy fiber-CA3 synapses in organotypic hippocampal slices prepared by high-pressure freezing and freeze substitution.

(A-D) Electron micrograph of a mossy fiber bouton and CA3 pyramidal neuron from a wild-type hippocampal slice culture at DIV14. Arrowheads highlight the three contact types mossy fiber boutons make with CA3 pyramidal neurons. White arrowheads indicate synaptic contact with spines, higher magnification in (B). Dark gray arrowheads mark the synaptic contacts onto the dendritic shaft of the CA3 pyramidal neuron, higher magnification in (C). Black arrowheads indicate a *puncta adherens* formed between a mossy fiber bouton and the dendritic shaft of a CA3 pyramidal neuron (D). Abbreviations: m, mitochondria; a, astrocytic process. Scale bars: 1 µm, A; 200 µm, B-D.

and *puncta adherens* onto the dendritic shaft (Figure 6 D). Synaptic contacts were characterized by the presence of a postsynaptic density, the widening of the synaptic cleft, and an accumulation of presynaptic vesicles in direct apposition to the postsynaptic density (Gray, 1959; Palay, 1956; Studer et al., 2014). *Puncta adherentia* contacts were distinguished by the presence of pre- and postsynaptic membrane specializations and a distinct absence of membrane-proximal presynaptic vesicles (Chicurel and Harris, 1992).

The ultrastructural morphology of large mossy fiber boutons in the *stratum lucidum* of cryo-fixed hippocampal slice cultures at DIV14 and mice transcardially perfused at P28 were found to be highly comparable (Figure A, B). The mossy fiber boutons in slices at DIV28 were also

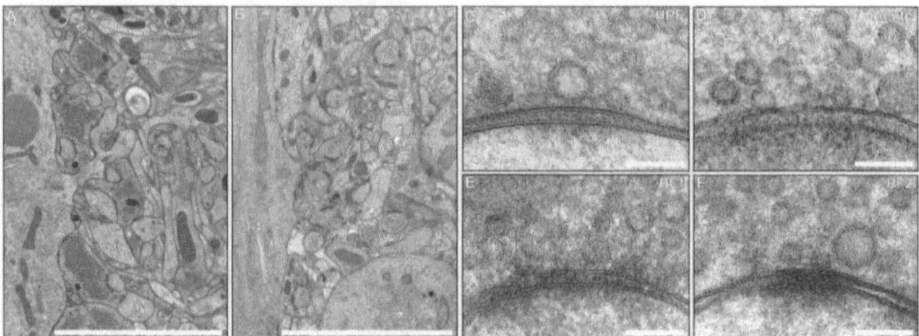

Figure 7. Comparative ultrastructural analysis of mossy fiber-CA3 synapse morphology in organotypic and *ex vivo* hippocampal preparations.

(**A, B**) Electron micrographs in the CA3 *stratum lucidum* from a hippocampal slice culture (**A**) and perfusion-fixed wild-type mouse hippocampus (**B**) with mossy fiber boutons engulfing complex spines of CA3 pyramidal neurons (false colored in yellow). (**C-F**) High magnification electron micrographs of mossy fiber-CA3 pyramidal neuron spine synapses from organotypic slice culture (**C**), acute slice preparation (**D**), and perfusion fixation (**E**) from Perfusion Fixative 1 (PF1; ice-cold 4% PFA, 2.5% glutaraldehyde in 0.1 M PB; and (**F**) from Perfusion Fixative 2 (PF2; 37° C, 2% PFA, 2.5% glutaraldehyde, 2 mM CaCl$_2$ in 0.1 M cacodylate buffer). Scale bars: 5 µm, **A** and **B**; 100 nm, **C-F**.

comparable to the perfused mossy fiber boutons at P28 (data not shown). A closer examination of individual mossy fiber active zone release sites in high-magnification electron micrographs revealed general structural similarities in high-pressure frozen and freeze-substituted slice cultures (Figure 7 C), acute slice preparations (Figure 7 D), and perfusion-fixed hippocampi based on two seminal ultrastructural studies; Perfusion Fixative 1 (PF1; Figure 7 E; ice cold 4% PFA, 2.5% GA in 0.1 M PB; Rollenhagen et al., 2007) and Perfusion Fixative 2 (PF2; Figure 7 F; 37°C, 2% PFA, 2.5% GA, 2 mM CaCl$_2$ in 0.1 M cacodylate buffer; Chicurel and Harris, 1992).

In conclusion, my data demonstrate that an *in vivo*-like anatomical organization of the hippocampal mossy fiber projection is retained in organotypic slice cultures and that the structural characteristics of the pre- and postsynaptic compartments of the mossy fiber-CA3 synapse are exquisitely preserved for ultrastructural analysis using a combination of HPF and freeze substitution.

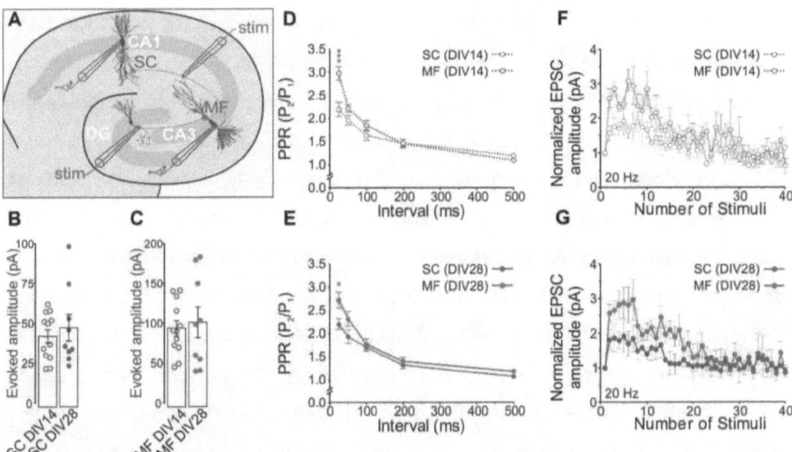

Figure 8. Comparison of Schaffer collateral and mossy fiber release probability and short-term plasticity characteristics from hippocampal slice cultures at DIV14 and 28.

(**A**) Schematic of electrophysiological experiments performed on hippocampal slice cultures, in red the placement of the stimulation electrode (stim) and the patch-clamped postsynaptic cell for mossy fiber recordings, and in blue the placement of the stimulation electrode (stim) and the patch-clamped postsynaptic cell for Schaffer collateral recordings. (**B**) Scatter plot of the mean evoked amplitude in CA1 pyramidal neurons from Schaffer collateral synapses slice cultures at DIV14 (open circles) and DIV28 (solid circles). (**C**) Scatter plot of the mean evoked amplitude in CA3 pyramidal neurons from mossy fiber synapses from slice cultures at DIV14 (open circles) and DIV28 (closed circles). (**D-E**) Paired-pulse ratios of Schaffer collateral (blue) and mossy fiber (red) synaptic responses from slice cultures at DIV14 (**D**) and DIV28 (**E**). (**F-G**) Normalized EPSC amplitude recorded in CA1 (blue) and CA3 (red) pyramidal neurons during a 20 Hz train of stimulation of Schaffer collateral and mossy fiber synapses, respectively, from slice cultures at DIV14 (**F**) and DIV28 (**G**). Abbreviations: DG, dentate gyrus; MF, mossy fiber; SC, Schaffer collateral; PPR, paired-pulse ratio. Statistical significance is represented as *, p<0.05; **, p<0.005; ***, p<0.001. N=3 cultures; n=12 cells SC at DIV14; 9 cells SC at DIV28; 11 cells MF DIV14; 9 cells MF DIV28. See Table 15 for all statistical analyses.

3.1.2. Comparative functional analysis of Schaffer collateral and mossy fiber synapses in hippocampal slice cultures

To assess whether the functionally well-characterized neurotransmitter release properties of Schaffer collateral and mossy fiber-CA3 synapses described in acute hippocampal slice preparations are also preserved in cultured organotypic slices, electrophysiological recordings of postsynaptic responses were made from CA1 and CA3 pyramidal neurons during fiber stimulation of the Schaffer collateral and mossy fiber axon projections, respectively (Figure 8) performed by Chungku Lee). Whole-cell voltage-clamp recordings from CA1 pyramidal neurons were performed during fiber stimulation of Schaffer collateral axons (Figure 8 A, blue). No difference in the evoked EPSC amplitude was found between DIV14 and DIV28 (Figure 8 B; 42.23 ± 4.252 pA, SC DIV14; 47.64 ± 8.251 pA, SC DIV28; p=0.54). Similarly, measurements from CA3 pyramidal neurons in response to stimulation of the mossy fiber axonal pathway in the hilus (Figure 8 A, red) showed no significant difference in the evoked EPSC amplitude between the two developmental time-points (Figure 8 C; 93.96 ± 9.533 pA, MF DIV14; 101.3 ± 19.12 pA, MF DIV28; p=0.71). The paired-pulse ratio for mossy fiber synapses was significantly higher in response to closely spaced stimuli (25 ms interstimulus interval) compared to Schaffer collateral synapses at both DIV14 (Figure 8 D; 2.203 ± 0.116, SC DIV14; 2.981 ± 0.145, MF DIV14; p<0.001) and DIV28 (Figure 8 E; 2.195 ± 0.160, SC DIV28; 2.712 ± 0.169, MF DIV28; p=0.04). Correspondingly, mossy fiber synapses demonstrated a considerably higher degree of facilitation than Schaffer collaterals in response to high-frequency fiber stimulation trains (Figure 8 F and G) (40 stimuli delivered at 20 Hz). These synapse-specific differences in short-term plasticity were similarly observed at both DIV14 (Figure 8 F) and DIV28 (Figure 8 G) developmental time-points. Taken together, these data indicate that fundamental and presynaptically expressed differences in release probability and short-term plasticity also exist between Schaffer collateral and mossy fiber-CA3 synapses in a cultured organotypic slice context.

3.1.3. Comparative ultrastructural analysis of Schaffer collateral and mossy fiber synapses in hippocampal slice cultures

To investigate whether the differences in functional release properties of Schaffer collateral and mossy fiber-CA3 synapses correlate with distinct ultrastructural profiles of synaptic vesicle organization at individual active zone release sites, I performed a comparative electron tomographic analysis of the respective synapse types. I compared Schaffer collateral and mossy fiber synapses from within the same organotypic slice to minimize experimental variability. Tomographically reconstructed active zone subvolumes from mossy fiber-CA3 (Figure 9 A) and Schaffer collateral (Figure 9 D) synapses enabled accurate resolution and quantification of docked synaptic vesicles (green arrowheads) in direct contact with the presynaptic membrane. Segmented 3D models generated from the tomographic subvolumes (Figure 9 B, C, E, F) allowed for the precise extraction of the spatial coordinates of synaptic vesicles with respect to the relative position and surface area of reconstructed active zones. This permitted numbers of docked synaptic vesicles to be normalized to reconstructed active zone areas (Figure 9 C, F), thereby controlling for variations in section thickness or the size of reconstructed active zones.

I first analyzed the spatial distribution of vesicles with respect to the active zone membrane. In both mossy fiber-CA3 and Schaffer collateral synapses, the majority of all vesicles within 100 nm of the active zone were positioned within 0-2 nm of the plasma membrane and therefore "docked" at the active zone (Figure 9 G). However, the number of docked vesicles normalized to the active zone area was significantly lower in mossy fiber synapses compared to Schaffer collateral synapses (Figure 9 H; 0-2 nm: 1.015 ± 0.112, SC; 0.576 ± 0.576; MF; p=0.003). This synapse-specific difference in the abundance of membrane-proximal vesicles was also observed for vesicles within 0-5 nm of the active zone membrane (Figure 9 I; 0-5 nm: 1.149 ± 0.127, SC; 0.687 ± 0.095, MF; p=0.006). Interestingly, mossy fiber-CA3 synapses exhibited a second membrane-proximal pool of vesicles within 5-20 nm of the active zone membrane that was absent in Schaffer collateral synapses (Figure 9 I; 5-10 nm: 0.239 ± 0.054, SC; 0.471 ± 0.076, MF; p=0.005). The average active zone size sampled in reconstructed synaptic subvolumes of Schaffer collateral synapses was significantly smaller than in mossy fiber synapses (Figure 9J; 3.495 ± 0.185, SC; 4.470 ± 0.260 0.01 µm², MF; p=0.003). A lower proportion of vesicles within 40 nm of the active zone was morphologically docked in mossy fiber compared to Schaffer collateral synapses (Figure 9 K; 0.471 ± 0.040, SC; 0.297 ± 0.038, MF; p=0.002). Despite such differences in their relative distribution, the number of vesicles within 40 nm of the active zone (Figure 9 L; 2.137 ± 0.137, SC; 2.072 ± 0.141, MF; p=0.74), and within 100 nm of the active zone were highly comparable between mossy fiber and Schaffer collateral synapses at DIV14 (Figure 9 M; 6.534 ± 0.276, SC; 6.356 ± 0.415, MF; p=0.71).

I similarly analyzed the distribution of vesicles at active zones from Schaffer collateral and mossy fiber synapses at DIV28 to investigate potential developmental changes in synaptic vesicle organization at the active zone. In comparison to Schaffer collateral and mossy fiber synapses at DIV14, both Schaffer collateral and mossy fiber synapses at DIV28 exhibited an increase in docked synaptic vesicles (Figure 9 N and O). However, mossy fiber synapses persistently harbored a lower spatial density of docked vesicles (Figure 9 O; 1.710 ± 0.139, SC; 1.061 ± 0.141, MF; p<0.001) and vesicles within 0-5 nm of the active zone (Figure 9 P; 1.721 ± 0.138, SC; 1.071 ± 0.124, MF; p<0.001) as well as the prominent accumulation of membrane-proximal vesicles (Figure 9 P; 5-10 nm: 0.072 ± 0.024, SC; 0.332 ± 0.055, MF; p<0.001; 10-

20 nm: 0.100 ± 0.033, SC; 0.323 ± 0.051, MF; p<0.001; 20-30 nm: 0.109 ± 0.027, SC; 0.242 ± 0.047, MF; p=0.03) compared to Schaffer collateral synapses. Unlike synaptic subvolumes at DIV14, the active zone area sampled at DIV28 was similar in both Schaffer collateral and mossy fiber synapses (Figure 9 Q; 4.558 ± 0.279, SC; 5.533 ± 0.400 0.01 µm², MF; p=0.08). Consistent with my previous observation, Schaffer collateral synapses maintained a comparatively higher proportion of docked vesicles relative to all vesicles within 40 nm of the active zone (Figure 9 R; 0.724 ± 0.032, SC; 0.452 ± 0.042; p<0.001). The total number of vesicles within 40 nm (Figure 9 S; 2.312 ± 0.148, SC; 2.300 ± 0.164, MF; p=0.95) and within 100 nm of the active zone was highly comparable between mossy fiber and Schaffer collateral synapses at DIV28 (Figure 9 T; 6.478 ± 0.347, SC; 6.193 ± 0.470, MF; p=0.63).

In conclusion, my data reveal that mossy fiber active zones harbor fewer docked vesicles than Schaffer collaterals at both DIV14 and DIV28. Interestingly, mossy fiber synapses are also characterized by a prominent membrane-proximal pool of vesicles located 5-20 nm from the active zone membrane that is not observed in wild-type Schaffer collateral synapses. This comparative disparity in vesicle organization is indicative of different spatial distributions rather than vesicle availability, since the total number of vesicles within 40 nm or 100 nm of the active zone are highly comparable in both Schaffer collateral and mossy fiber synapses.

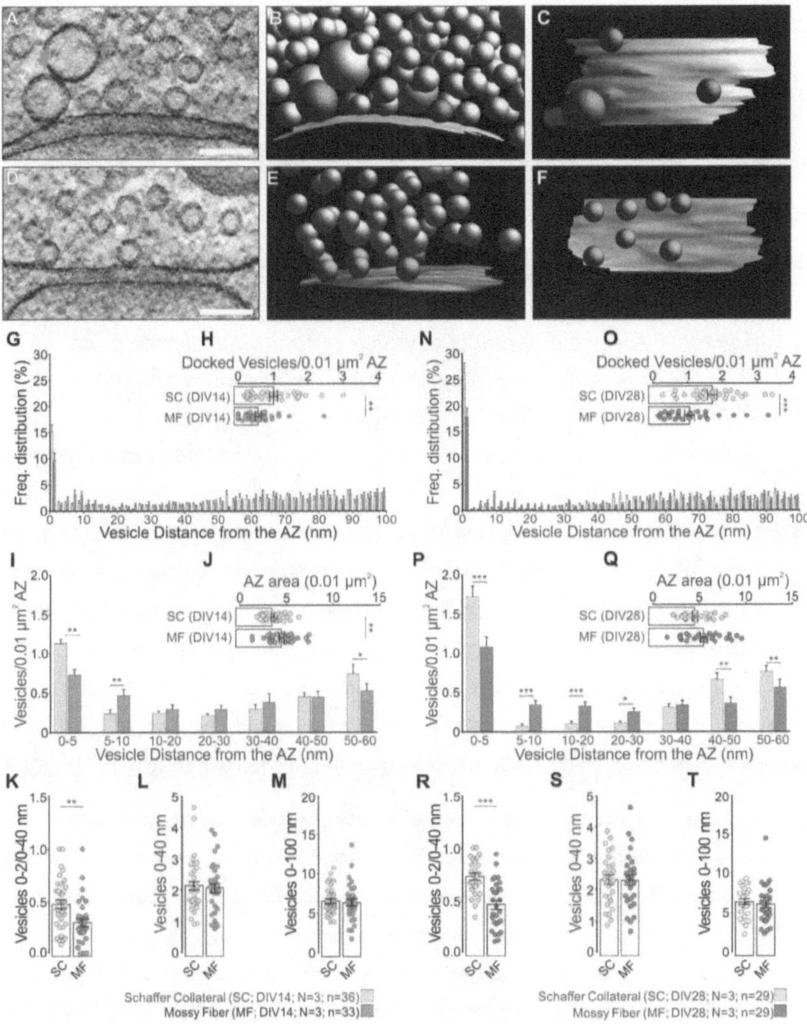

Figure 9. Comparative ultrastructural analysis of Schaffer collateral and mossy fiber active zones in organotypic hippocampal slice slices at DIV14 and DIV28.

(A-M) Analysis of Schaffer collateral and mossy fiber synapses from hippocampal slice cultures at DIV14. (N-T) Analysis of Schaffer collateral and mossy fiber synapses from hippocampal slice cultures at DIV28. (A, D) Tomographic subvolumes of mossy fiber (A) and Schaffer collateral (D) active zones (docked synaptic vesicles, green arrowheads). (B, E) Corresponding 3D models of mossy fiber (B) and Schaffer collateral (E) synapses (docked synaptic vesicles, green spheres; non-docked synaptic vesicles, gray spheres; active zone,

61

(Figure 9. continued): active zone, gray). (**C, F**) Orthogonal views of the reconstructed mossy fiber (**C**) and Schaffer collateral (**F**) active zones and the spatial distribution of docked synaptic vesicles. (**G, N**) Frequency distribution of vesicles within 100 nm of the active zone membrane. (**H, O**) Scatter plots of docked vesicles (within 0-2 nm) at the active zone. (**I, P**) Spatial distribution of vesicles in 5 and 10 nm bins from the active zone normalized to active zone area. (**J, Q**) Scatter plots of active zone areas. (**K, R**) Scatter plots of the relative proportion of docked vesicles of all vesicles within 40 nm of the active zone in Schaffer collateral and mossy fiber synapses. (**L, M, S, T**) Scatter plots of the number of synaptic vesicles within 40 nm (**L, S**) and 100 nm (**M, T**) of the active zone normalized to active zone area. Statistical significance is represented as *, p<0.05; **, p<0.005; ***, p<0.001. N= number of cultures; n=number of active zones. Scale bars: 100 nm, **A-F**. See Tables 15-18 for all statistical analyses for DIV14 and DIV28, respectively.

3.1.4. Synapse-specific differences in vesicle organization characterize Schaffer collateral and mossy fiber synapses in high-pressure frozen acute hippocampal slice preparations

To investigate whether the difference in the density of docked vesicles at individual active zones between mossy fiber and Schaffer collateral synapses are similarly manifest in synapses which developed *in vivo*, I compared the ultrastructural organization of synaptic vesicle pools in Schaffer collateral and mossy fiber synapses in acute slice preparations age-matched to DIV14 slice cultures. To account for the fact that hippocampal slice cultures were prepared from mouse pups at P3-7, I prepared acute hippocampal slices from P18 wild-type mice for HPF cryofixation according to a published protocol (Korogod et al., 2015) and analyzed Schaffer collateral (Figure 10 A) and mossy fiber synapses (Figure 10 B) using 3D reconstructions of the synaptic subvolumes imaged with electron tomography.

As in organotypic slice cultures, the majority of all vesicles within 100 nm of Schaffer collateral and mossy fiber synapses were docked (within 0-2 nm of the active zone membrane) (Figure 10 C). Moreover, fewer docked synaptic vesicles were observed at mossy fiber synapses than at Schaffer collateral synapses when normalized to active zone area (Figure 10 D; 1.608 ± 0.163, SC; 1.128 ± 0.095, MF; p=0.009), consistent with the findings *in vitro* (Figure 9 H). Although there was a trend of more vesicles within 10-20 nm of the active zone in mossy fiber synapses in comparison to Schaffer collateral synapses in acute preparations, this difference did not reach statistical significance (Figure 10 E; 10-20 nm: 0.302 ± 0.070, SC; 0.455 ± 0.061, MF; p=0.14). Consistent with Schaffer collateral and mossy fiber synapses from organotypic cultures, the active zone area sampled in Schaffer collateral synapses from acute slice preparations was smaller than mossy fiber active zones from the same slice (Figure 10 F; 3.979

± 0.226, SC; 5.538 ± 0.331 0.01 µm², MF; p=0.001). In mossy fiber synapses from acute preparations, a lower proportion of docked vesicles from all vesicles within 40 nm of the active zone was found in comparison to Schaffer collateral synapses (Figure 10 G; 0.617 ± 0.053, SC; 0.448 ± 0.037, MF; p=0.009). Both synapse types had comparable numbers of synaptic vesicles within 40 nm (Figure 10 H; 2.671 ± 0.211, SC; 2.566 ± 0.188, MF; p=0.72) and 100 nm (Figure 10 I; 6.873 ± 0.437, SC; 6.820 ± 0.488, MF; p=0.66) of the active zone membrane.

In summary, my findings indicate that mossy fiber synapses are characterized by a comparatively lower spatial density of docked synaptic vesicles in both *in vitro* and *in vivo* preparations, thus excluding the possibility that this observation is an artifact of the slice culture system.

3.1.5. Perfusion fixation of brain tissue causes a severe reduction in docked and membrane-proximal synaptic vesicles in mossy fiber-CA3 synapses

Two seminal ultrastructural studies of the mossy fiber-CA3 synapse used perfusion-fixed hippocampi from rats (Chicurel and Harris, 1992; Rollenhagen et al., 2007). To test whether perfusion of chemical fixatives changed the spatial distribution of synaptic vesicles at mossy fiber synapses, I perfusion-fixed wild-type mice at P28 following the protocols from these studies (Chicurel and Harris, 1992; Rollenhagen et al., 2007). This experiment served as another *ex vivo* preparation to assess the abundance and spatial distribution of synaptic vesicles using electron tomography. Post-fixed brains were vibratomed and regions of the CA3 were excised with a biopsy punch before being high-pressure frozen and processed by AFS (Möbius et al., 2010). This approach was designed to focus my investigation primarily on the potential effects of aldehyde fixation on vesicle organization, rather than on shrinkage artifacts introduced by classic room-temperature dehydration steps. I found that the general morphology of mossy fiber synapses in both of the perfusion-fixed tissue to be comparable to that observed in hippocampal slice culture (Figure 7).

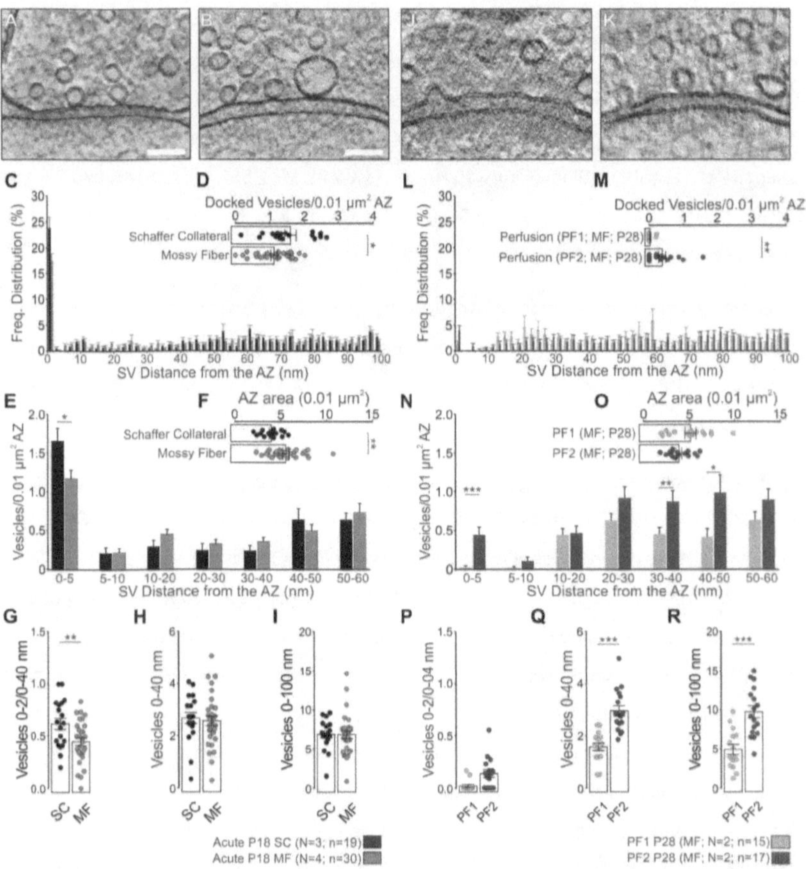

Figure 10. Ultrastructural analysis from *ex vivo* preparations of hippocampal tissue.

(A-I) Analysis of Schaffer collateral and mossy fiber synapses from acute hippocampal slice preparations from P18 wild-type mice. (J-R) Analysis of mossy fiber synapses from hippocampal tissue from P28 wild-type mice transcardially perfused with one of two aldehyde cocktails: PF1 (ice-cold 4% PFA, 2.5% GA in 0.1 M PB; or PF2 (37° C, 2% PFA, 2.5% GA, 2 mM $CaCl_2$ in 0.1 M cacodylate buffer). (A-B, J-K) Tomographic subvolumes from Schaffer collateral (A) and mossy fiber (B, J, K) synapses. (C, L) Frequency distributions of vesicles within 100 nm of the active zone membrane. (D, M) Scatter plots of docked vesicles (0-2 nm) at the active zone membrane normalized to active zone area. (E, N) Spatial distribution of vesicles in 5 and 10 nm bins from the active zone membrane normalized to active zone area. (F, O) Scatter plots of active zone areas. (G, P) Scatter plots of the relative proportion of docked vesicles of all vesicles within 40 nm of the active zone. (H, I, Q, R) Scatter plots of the number of synaptic vesicles within 40 nm (H, Q) and 100 nm (I, R) of the active zone normalized to active zone. Statistical significance is represented as *, $p<0.05$; **, $p<0.005$; ***, $p<0.001$. N= number of animals; n=number of active zones. Scale bars: 100 nm, A, B, J, and K. See Tables 19-22 for full statistical analyses for acute and perfusion-fixed experiments, respectively.

In reconstructed electron tomograms it became evident that the two fixation protocols used had stimulated vesicle fusion (Figure 10 J and K). The two perfusion protocols were as follows: i) PF1 comprised of ice-cold 4% PFA, 2.5% GA dissolved in in 0.1 M PB at a pH of 7.4 (Figure 10 J) (Rollenhagen et al., 2007), and ii) PF2 comprised of 2% PFA, 2.5% GA, 2 mM $CaCl_2$ dissolved in 0.1 M cacodylate buffer at a pH of 7.4 at 37° C (Figure 10 K) (Chicurel and Harris, 1992). The spatial distribution of synaptic vesicles was different to that observed in age-matched cryo-fixed slice cultures (Figure 10 L, perfusion fixation; Figure 9 N, cryo-fixed wild-type slice culture). There were fewer docked synaptic vesicles in both fixation protocols compared to vesicle docking in mossy fiber synapses from acute slice preparations (Figure 10 M; 0.027 ± 0.019, PF1; 0.394 ± 0.107, PF2; p=0.002; acute slice: Figure 10 D). In addition, the spatial density of synaptic vesicles in mossy fiber synapses from both perfusion fixation protocols lacked synaptic vesicles in proximity to the active zone (Figure 10 N; 0-5 nm: 0.027 ± 0.019, PF1; 0.486 ± 0.103, PF2; p<0.001; 5-10 nm: 0.021 ± 0.015, PF1; 0.139 ± 0.052, PF2; p=0.05). The density of synaptic vesicles increased beyond 20 nm from the active zone (Figure 10 N; 20-30 nm: 0.629 ± 0.090, PF1; 0.956 ± 0.143, PF2; p=0.09; 30-40 nm: 0.450 ± 0.089, PF1; 0.871 ± 0.137, PF2; p=0.007). This is likely due to a smaller active zone area measured in the PF2 condition compared to PF1 and age-matched cryo-fixed slice cultures (Figure 10 O; 5.122 ± 0.592 0.01 μm^2, PF1; 3.854 ± 0.307 0.01 μm^2, PF2; p=0.06). The proportion of docked vesicles from all vesicles within 40 nm was very low in both perfusion conditions (Figure 10 P 0.019 ± 0.013, PF1; 0.137 ± 0.036, PF2; p=0.003), which was far below the ratio observed in mossy fiber synapses from age-match wild type slice cultures (Figure 9 R; DIV28). There were, however, more synaptic vesicles in PF2 within 40 nm (Figure 10 Q; 1.573 ± 0.150, PF1; 2.954 ± 0.195, PF2; p<0.001) and 100 nm (Figure 10 R; 4.887 ± 0.651, PF1; 9.678 ± 0.754, PF2; p<0.001) than in PF1, likely as a result of the smaller active zone area samples in PF2, used for normalization.

These findings indicate that perfusion fixation of different aldehyde cocktails caused severe reductions in membrane-proximal vesicle pools in hippocampal mossy fiber synapses. The omega-shaped exo-endocytic intermediates observed in these conditions indicate that the observed depletion of membrane-attached and membrane-proximal vesicles is likely due to induced vesicle fusion.

3.1.6. Munc13 priming molecules are essential for vesicle docking in hippocampal mossy fiber synapses

Previous studies have demonstrated that Munc13 priming molecules are required for morphological docking of synaptic vesicles at small, glutamatergic spine synapses in the mouse hippocampus (Imig et al., 2014; Siksou et al., 2009a). These data, together with work in invertebrate model systems (Böhme et al., 2016; Hammarlund et al., 2007; Weimer et al., 2006), indicate that this molecular requirement is evolutionarily conserved. Nevertheless, the discovery that some functionally and structurally specialized synapses, including retinal photoreceptor ribbon synapses (Cooper et al., 2012) and cochlear hair cell ribbon synapses (Vogl et al., 2015) function in the absence of Munc13-mediated priming proteins, emphasizes the importance of investigating the molecular requirements for synapse function on a subtype specific basis. Although electrophysiological studies have investigated the roles of individual Munc13 isoforms in presynaptic forms of synaptic plasticity at the mossy fiber-CA3 synapse (Breustedt et al., 2010; Yang and Calakos, 2011), the role of Munc13 priming molecules in synaptic vesicle organization at these synapses remains unclear since electron microscopic analyses have thus lacked the resolution to accurately quantify membrane-attached vesicles (Zhao et al., 2012a, 2012b).

To examine the role of Munc13 proteins in synaptic vesicle docking in hippocampal mossy fiber synapses, I generated slice cultures from Munc13-1/2 DKO (Munc13-1$^{-/-}$, Munc13-2$^{-/-}$) and CTRL littermates (Munc13-1$^{+/-}$, Munc13-2$^{+/-}$). To circumvent problems associated with the perinatal lethality of the Munc13-1 constitutive KO (Augustin et al., 1999), slice cultures were prepared from E18 mouse pups. The gross morphology of Munc13-deficient hippocampal mossy fibers appeared highly comparable to controls and gave no indication of developmental deficits, consistent with previous studies (Augustin et al., 1999; Sigler et al., 2017). In line with past studies (Imig et al., 2014; Siksou et al., 2009a), an analysis of electron tomograms from Munc13-1/2 CTRL (Figure 11 A, B, F) and DKO (Figure 11 C, D, F) mossy fiber-CA3 synapses revealed a complete loss of vesicle docking as indicated by the scarcity of vesicles within both 0-2 nm (Figure 11 F; Munc13s, 0.861 ± 0.159, CTRL; 0 ± 0, DKO; p<0.001) and 0-5 nm (Figure 11 G; 0-5 nm: 0.923 ± 0.165, CTRL; 0 ± 0, DKO; p<0.001) of the active zone

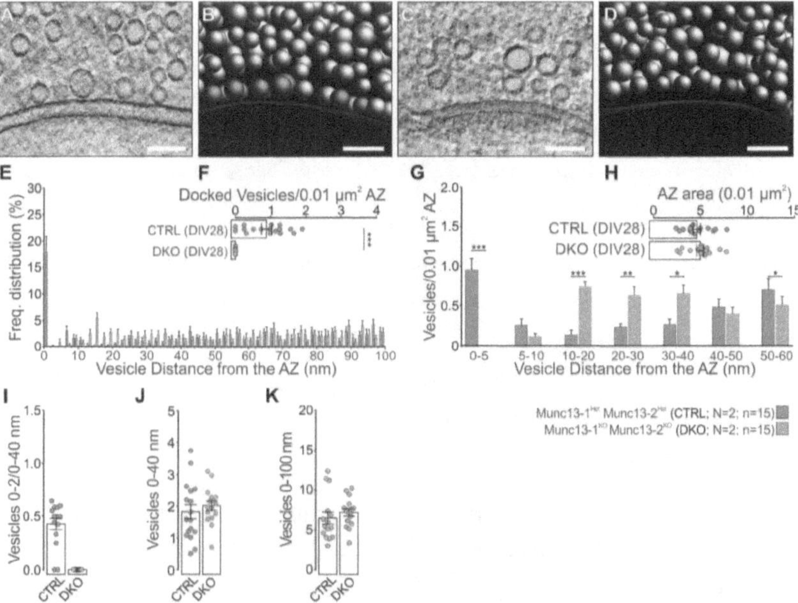

Figure 11. Ultrastructural analysis of mossy fiber active zones in Munc13-deficient and control slice cultures at DIV28.

(A, C) Tomographic subvolumes of mossy fiber synapses from Munc13-1/2 CTRL (A; Munc13-1$^{+/-}$ Munc13-2$^{+/-}$) and DKO (C; Munc13-1$^{-/-}$ Munc13-2$^{-/-}$) slice cultures. (B, D) Corresponding 3D models of CTRL (B) and DKO (D) mossy fiber synapses (docked synaptic vesicles, green; DCVs, orange; non-docked synaptic vesicles, gray; active zone, gray). (E) Frequency distribution of vesicles within 100 nm of the active zone. (F) Scatter plot of docked synaptic vesicles (0-2 nm) normalized to the active zone area. (G) Spatial distribution of vesicles in 5 and 10 nm bins normalized to active zone area. (H) Scatter plot of active zone areas. (I) Scatter plot of the relative proportion of docked vesicles from all vesicles within 40 nm of the active zone. (J, K) Scatter plots of the number of vesicles within 40 nm (J) and 100 nm (K) of the active zone normalized to active zone area. Scale bars: 100 nm, A-D. Statistical significance is represented as *, p<0.05; **, p<0.005; ***, p<0.001. N= number of cultures; n=number of active zones. See Table 23 and 24 for full statistical analysis.

membrane. Munc13-1/2 DKO mossy fiber synapses also exhibited a prominent accumulation of membrane-proximal vesicles between 10-40 nm from the active zone (Figure 11 G; 10-20 nm: 0.155 ± 0.068, CTRL; 0.762 ± 0.059, DKO; p<0.001; 20-30 nm: 0.224 ± 0.048, CTRL; 0.700 ± 0.115, DKO; p=0.002; 30-40 nm: 0.305 ± 0.067, CTRL; 0.633 ± 0.106, DKO; p=0.01). The active zone area sampled from CTRL and DKO synaptic subvolumes was the same (Figure 11 H; 4.615 ± 0.387, CTRL; 4.976 ± 0.379 0.01 μm², MF; p=0.51). In CTRL tomograms, the density of docked vesicles (Figure 11 F, CTRL; Figure 9 O, wild-type) and the proportion of docked vesicles from all vesicles within 40 nm of the active zone (Figure 11 I, CTRL, and Figure 9 R, wild-type) were highly comparable to wild-type mossy fiber synapses at DIV28. Despite severe docking deficits, vesicle recruitment within 40 nm (Figure 11 J; 1.903 ± 0.253, CTRL; 2.303 ± 0.156, DKO; p=0.19) and 100 nm (Figure 11 K; 5.983 ± 0.728, CTRL; 6.499 ± 0.461, DKO; p=0.55) of the active zone was unaffected in Munc13-1/2 DKO mossy fiber synapses.

My findings indicate that Munc13s are essential for the morphological docking of vesicles at the mossy fiber synapses. Despite the severe docking deficit in Munc13-1/2 DKO mossy fiber synapses, the number of vesicles within 40 nm of the active zone was the same as in CTRL mossy fiber synapses. Furthermore, the density of docked vesicles in Munc13 CTRL mossy fiber synapses was the same as in wild-type mossy fiber synapses at DIV28. Despite the heterozygous genotype of Munc13-1 and -2, this was enough to achieve wild-type-like vesicle docking in mossy fiber synapses.

3.1.7. Acute pharmacological inhibition of action potential firing does not alter docked synaptic vesicle density in mossy fiber synapses

To address whether endogenous synaptic transmission from mossy fiber synapses influenced the spatial distribution of synaptic vesicles at the active zone, activity was blocked in wild-type hippocampal slices by an acute pharmacological treatment prior to high-pressure freeze fixation. Slices in this experiment were exposed to one of three conditions: (i) a VC, in which slices were treated with their own medium; (ii) T/N/A: a cocktail designed to block slice activity, containing TTX (T) to prevent sodium-propagated action potentials (Narahashi et al., 1964), and the postsynaptic glutamate receptor blockers NBQX (N) and D-AP5 (A) to inhibit AMPA and NMDA receptors, respectively; and (iii) T/D: a cocktail designed to selectively suppress mossy fiber synaptic transmission, comprising TTX (T) with the mGluR2 receptor agonist DCG-IV (D), which reduces synaptic transmission from mossy fiber synapses via reduction of intracellular cAMP levels (Kamiya and Ozawa, 1999; Kamiya et al., 1996). The spatial distribution of synaptic vesicles within 100 nm of mossy fiber active zones was unchanged after acute pharmacological silencing of network activity with both T/N/A and T/D treatment compared to VC (Figure 12 A). The density of docked vesicles was comparable between VC and pharmacologically silenced mossy fiber synapses (Figure 12 B; 0.634 ± 0.110, VC; 0.605 ± 0.066, T/N/A; 0.688 ± 0.101, T/D; p=0.82). In synapses treated with T/D, vesicle numbers were increased within 0-5 nm (Figure 12 C; 0-5 nm: 0.734 ± 0.119, VC; 0.758 ± 0.077,

T/N/A; 0.946 ± 0.118, T/D; p=0.31), 5-10 nm (Figure 12 C; 5-10 nm: 0.519 ± 0.112, VC; 0.531 ± 0.090, T/N/A; 0.689 ± 0.080, T/D; p=0.08), and 10-20 nm (Figure 12 C; 10-20 nm: 0.411 ± 0.109, VC; 0.333 ± 0.071, T/N/A; 0.483 ± 0.082, T/D; p=0.33) from the active zone, however, this trend did not reach statistical significance. The active zone area measured in synaptic subvolumes from the three treatment conditions was comparable (Figure 12 D; 3.506 ± 0.252, VC; 3.752 ± 0.259, T/N/A; 3.964 ± 0.295 0.01 μm^2, T/D; p=0.52). Despite the mild increase in membrane-proximal vesicles after acute application of T/D, the proportion of docked vesicles to all vesicles within 40 nm of the mossy fiber active zone was highly comparable to T/N/A-treated mossy fiber synapses and VCs (Figure 12 E; 0.265 ± 0.051, VC; 0.261 ± 0.030, T/N/A; 0.269 ± 0.032, T/D; p=0.99). There was no change in the density of synaptic vesicles within 40 nm (Figure 12 F; 2.318 ± 0.261, VC; 2.389 ± 0.181, T/N/A; 2.622 ± 0.142, T/D; p=0.12) or 100 nm (Figure 12 G; 6.763 ± 0.609, VC; 6.868 ± 0.416, T/N/A; 6.390 ± 0.419, T/D; p=0.62) after acute pharmacological blockade of network activity and of mossy fiber synaptic transmission when compared to control. As expected, mossy fiber synapses exhibited a tendency towards fewer morphological fusion events after T/N/A and T/D treatments, although this reduction did not reach statistical significance with the sample size of this study (data not shown; 0.064 ± 0.027, VC; 0.046 ± 0.020, T/N/A; 0.016 ± 0.011, T/D; p=0.34).

Taken together, these data indicate that the spatial organization of synaptic vesicles at mossy fiber active zones are rather insensitive to acute manipulations of spontaneous slice activity. The comparatively lower numbers of docked vesicles observed in mossy fiber synapses compared to Schaffer collateral synapses is therefore unlikely to result from the potential hyper-excitability of the mossy fiber pathway in slice cultures.

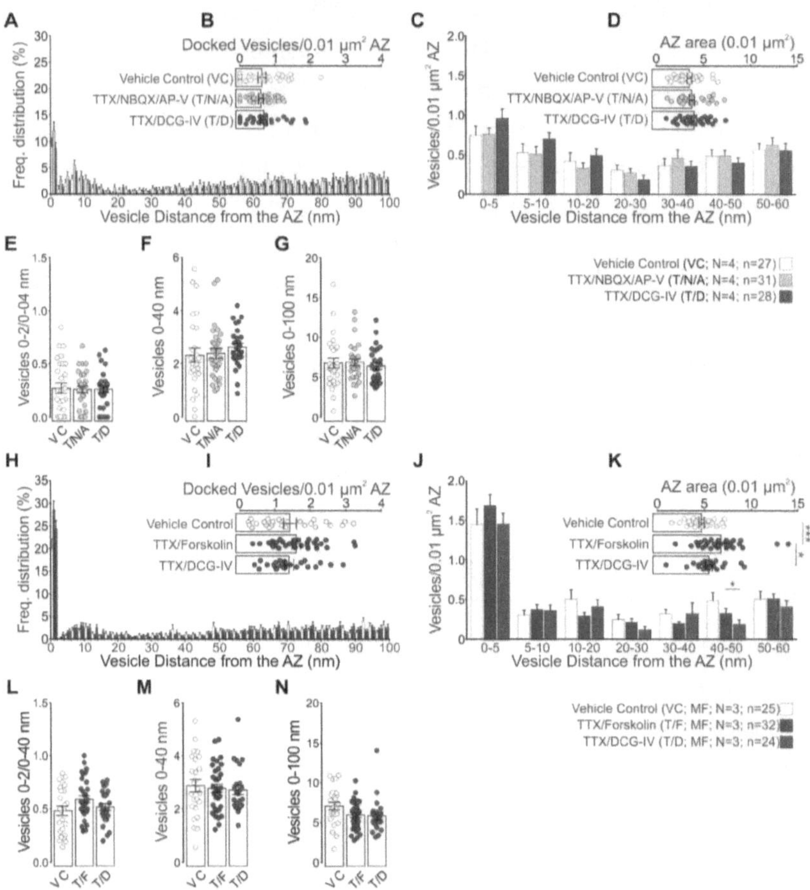

Figure 12. Ultrastructural analysis of vesicle pools after acute pharmacological manipulations of mossy fiber synapses in organotypic slice cultures at DIV14 and DIV28.

(**A-G**) Analysis of mossy fiber synapses after a 10 minute acute pharmacological silencing of slice cultures at DIV14 with one of three conditions: 1) T/N/A: TTX, to block sodium-propagated action potentials, supplemented with either NBQX and D-AP5 to block AMPA and NMDA receptor blockers, respectively; 2) T/D: TTX with DCG-IV, an mGluR2 receptor agonist that specifically blocks synaptic transmission in mossy fiber synapses (Kamiya and Ozawa, 1999); and 3) VC: comprised of slice culture medium. (**H-N**) Analysis of mossy fiber synapses from hippocampal slice cultures at DIV28 after a 15 minute treatment with either: 1) T/F: TTX, to block sodium propagated action potentials, and forskolin, an activator of AC1 which increases presynaptic cAMP concentrations (Chavez-Noriega and Stevens, 1994; Dixon and Atwood, 1989; Seamon et al., 1983) thus increasing synaptic transmission (Evans and Morgan, 2003; Seino and Shibasaki, 2005;

(**Figure 12**. continued): Weisskopf et al., 1994); 2) T/D: TTX with DCG-IV, and mGluR2 receptor agonist that specifically blocks synaptic transmission in mossy fiber synapses (Kamiya and Ozawa, 1999); and 3) VC: comprised of TTX in slice culture medium. (**A, H**) Frequency distributions of vesicles within 100 nm of the active zone membrane. (**B, I**) Scatter plots of docked vesicles (0-2 nm) at the active zone membrane normalized to active zone area. (**C, J**) Spatial distribution of vesicles in 5 and 10 nm bins normalized to active zone area. (**D, K**) Scatter plots of active zone area. (**E, L**) Scatter plots of the relative proportion of docked vesicles within 40 nm of the active zone. (**F, G, M, N**) Scatter plots of the number of synaptic vesicles within 40 nm (**F, M**) and 100 nm (**G, N**) of the active zone normalized to active zone area. Statistical significance is represented as *, p<0.05; **, p<0.005; ***, p<0.001. N= number of cultures; n=number of active zones. See Tables 25-26 and 27-28 for full statistical analyses of acutely silenced and acute pharmacological manipulation of cAMP experiments, respectively.

3.1.8. Pharmacological manipulation of presynaptic cAMP only minimally impacts synaptic vesicle organization in mossy fiber synapses

To investigate whether induced changes in presynaptic cAMP concentrations are reflected by corresponding changes in synaptic vesicle organization at mossy fiber active zones, I exploited the well-studied presynaptic cAMP-dependence of mossy fiber synaptic release probability (Evans and Morgan, 2003; Seino and Shibasaki, 2005; Weisskopf et al., 1994). I treated wild-type slice cultures at DIV28 with drug cocktails to either increase or decrease presynaptic cAMP concentrations. Three conditions were analyzed in this experiment: 1) T/F: TTX, to block sodium propagated action potentials, and forskolin, an activator of AC1 which increases presynaptic cAMP concentrations (Chavez-Noriega and Stevens, 1994; Dixon and Atwood, 1989; Seamon et al., 1983), thus increasing synaptic transmission (Evans and Morgan, 2003; Seino and Shibasaki, 2005; Weisskopf et al., 1994); 2) T/D: TTX with DCG-IV, an mGluR2 receptor agonist that specifically blocks synaptic transmission in mossy fiber synapses (Kamiya and Ozawa, 1999); and 3) VC: comprised of TTX in slice culture medium. Both forskolin and DCG-IV were effective in modulating spontaneous transmitter release in mossy fiber synapses in a 15-minute time frame, which was validated via electrophysiology (recordings by Bekir Altas from DIV14 slice cultures; data not shown). Slice cultures were then fixed and processed accordingly: high-pressure freeze-fixation, AFS, plastic embedding, and 3D electron tomography. The spatial distribution of synaptic vesicles within 100 nm of the active zone was comparable between the three pharmacological treatments (Figure 12 H). After increasing presynaptic mossy fiber cAMP levels with forskolin, there was a mild increase in docked synaptic vesicle density, although this tendency did not reach statistical significance (Figure 12 I; 1.422 ± 0.183, VC; 1.601 ± 0.118, T/F; 1.399 ± 0.129, T/D; p=0.40). Similarly, forskolin-treated mossy fiber synapses exhibited a mild increase in synaptic vesicles within 0-

5 nm of the active zone, normalized to the active zone area, however this tendency also did not reach statistical significance (Figure 12 J; 0-5 nm: 1.456 ± 0.191, VC; 1.692 ± 0.135, T/F; 1.463 ± 0.127, T/F; p=0.43). The second pool of membrane-proximal synaptic vesicles was unaffected by manipulation of presynaptic cAMP (Figure 12 H and J). There was an increase in the proportion of docked synaptic vesicles from all vesicles within 40 nm of the active zone after increasing presynaptic cAMP levels with forskolin that was reminiscent to the proportion observed at untreated Schaffer collateral synapses at DIV28 (Figure 12 L, 0.482 ± 0.044, VC; 0.589 ± 0.031, T/F; 0.517 ± 0.033, T/D; Figure 9 R, untreated DIV28 Schaffer collateral). The number of synaptic vesicles within 40 nm (Figure 12 M; 2.871 ± 0.233, VC; 2.775 ± 0.151, T/F; 2.713 ± 0.178, T/D; p=0.85) and 100 nm of the active zone membrane (Figure 12 N; 7.047 ± 0.473, VC; 5.912 ± 0.299, T/F; 5.829 ± 0.438, T/D; p=0.03) was unchanged after manipulation of presynaptic cAMP in mossy fiber synapses.

These findings indicate that acute pharmacological manipulation of presynaptic cAMP concentrations in mossy fiber synapses does not have a major impact on synaptic vesicle organization at the active zone. Rather, my data indicate that forskolin-induced enhancement of synaptic transmission acts via molecular mechanisms operating downstream of synaptic vesicle docking and priming.

3.2. Morphological heterogeneity of the docked vesicle pool in mossy fiber synapses

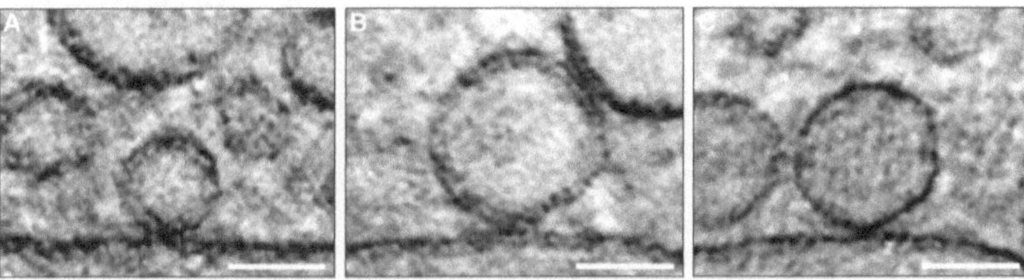

Figure 13. Three morphologically distinct vesicles dock at hippocampal mossy fiber synapses.

(**A-C**) Tomographic subvolumes of mossy fiber synapses showing docked synaptic vesicles (**A**), giant vesicles (**B**; clear-core vesicles with a diameter greater than 60 nm), and DCVs (**C**) Scale bars: 50 nm, **A-C**.

The second aim of my study was to perform a systematic analysis of the morphological nature of vesicles with respect to their subsynaptic distribution within Schaffer collateral and mossy fiber synapses. I was primarily motivated in this aim by the unexpected finding that mossy fiber active zones harbor a remarkable heterogeneity in vesicle size and morphological features. Of particular importance, my data reveal that docked vesicle pools at mossy fiber active zones comprise three morphological classes: 1) small, clear-cored synaptic vesicles with diameters ranging from 38 to 59 nm (Figure 13 A); 2) large diameter clear-core vesicles ("giant vesicles") with diameters ranging from 60 to 120 nm (Figure 13 B); and 3) DCVs with diameters ranging from 46 to 90 nm (Figure 13 C). In this section, I quantitate this heterogeneity and in particular, dissect ultrastructure-function relationships in hippocampal mossy fibers by relating the size and nature of morphologically distinct vesicle pools observed by electron tomography with functional pool measurements obtained via direct presynaptic capacitance recordings (Hallermann et al., 2003; Midorikawa and Sakaba, 2017).

3.2.1. Large, clear-core vesicles dock exclusively at mossy fiber active zones in hippocampal slice cultures

In striking contrast to Schaffer collateral synapses, mossy fiber synapses harbored an abundance of large diameter vesicles within reconstructed tomograms acquired from hippocampal slice cultures at DIV14 (Figure 14 A-D; Figure 23 A and B) and DIV28 (Figure 14 E-H; Figure 23 D and C). Large diameter (Ø>60 nm) vesicles were only very rarely observed in

Schaffer collaterals and were never detected within 60 nm of the active zone membrane (Figure 14 A, E, I). At both developmental time points, giant vesicles accumulated in proximity to mossy fiber active zone release sites and dock at the presynaptic membrane (Figure 14 A, E). Consequently, the mean diameter of vesicles docked at mossy fiber active zones was significantly larger than for those docked at Schaffer collaterals at DIV14 (Figure 14 B; 43.32 ± 0.354 nm, SC; 53.96 ± 1.859 nm, MF; p<0.001) and DIV28 (Figure 14 F; 44.37 ± 0.230 nm, SC; 49.40 ± 1.034 nm, MF; p<0.001). Giant vesicles comprised approximately 22% and 12% of the docked vesicle pool at mossy fiber active zones at DIV14 (Figure 14 C; 0 ± 0, SC; 22.47 ± 5.635%, MF; p<0.001) and DIV28 (Figure 14 G; 0 ± 0, SC; 11.67 ± 4.251%, MF; p<0.001), respectively. Even when excluding all giant vesicles, the average diameter of docked synaptic vesicles (Ø<60 nm) in mossy fiber synapses was larger than in Schaffer collateral synapses (DIV14; Table 16; 43.32 ± 0.35 nm, SC; 46.78 ± 0.46 nm, MF; p<0.001; DIV28; Table 18; 44.37 ± 0.23 nm, SC; 45.77 ± 0.37 nm, MF; p<0.001). The abundance of giant vesicles within 40 nm of mossy fiber active zones was also slightly reduced from DIV14 (Figure 14 D, 0.355 ± 0.055, MF) to DIV28 (Figure 14 H, 0.248 ± 0.055, MF). This change in the proportion of docked giant vesicles is likely due to an increase in the number of docked synaptic vesicles rather than a decrease in the number of giant vesicles (Figure 14 G, Table 18, DIV28; Figure 9 C, Table 16, DIV14).

My findings provide the first clear evidence that giant vesicles dock in physical contact with the active zone membrane and comprise a substantial proportion of the docked vesicle pool in mossy fiber synapses.

3.2.2. Mossy fiber giant vesicles are not a consequence of the slice culture procedure

To investigate the possibility that giant vesicles observed in mossy fiber synapses in organotypic hippocampal slices represent structural artifacts resulting from the slice culture procedure, I performed a comparative analysis of Schaffer collateral and mossy fiber vesicle organization in acute slices prepared from P18 wild-type mice (Figure 14 I-L). Consistent with my observations in cultured slices, giant vesicles were observed almost exclusively in mossy fiber synapses (Figure 14 I; Figure 23 E and F). Although giant vesicles were less abundant in tomograms from acute slices (Table 19) compared to age-matched cultured slices (Table 15), they still exhibited a clear tendency to accumulate in proximity to, and to dock at, mossy fiber active zone release sites (Figure 14 I). As in cultured slices (Figure 14 B, F), the mean diameter of docked vesicles in acute slices was larger in mossy fiber synapses compared to Schaffer collaterals (Figure 14 J; 45.59 ± 0.318 nm, SC; 50.67 ± 0.842 nm, MF; p<0.001). Giant vesicles comprised approximately 5% of the total docked vesicle population at mossy fiber active zones (Figure 14 K; 5.366 ± 1.595%, MF). Only one Schaffer collateral synaptic profile contained a docked giant vesicle. The majority of giant vesicles observed in mossy fiber synaptic profiles from acute slices accumulated within 40 nm of the active zone (Figure 14 L, acute slice, 0.012 ± 0.12, SC; 0.122 ± 0.024, MF; p=0.001).

These results indicate that giant vesicles are also a distinguishing ultrastructural feature of mossy fiber synapses *in vivo* and are capable of docking at the active zone membrane. Thus, giant vesicles are not merely an artifact of the organotypic slice culture procedure. Although giant vesicles appeared less abundant in acute compared to cultured slices, the potential consequences of mechanical trauma and anoxia induced during acute slice preparation on vesicle organization must be taken into consideration when interpreting this result.

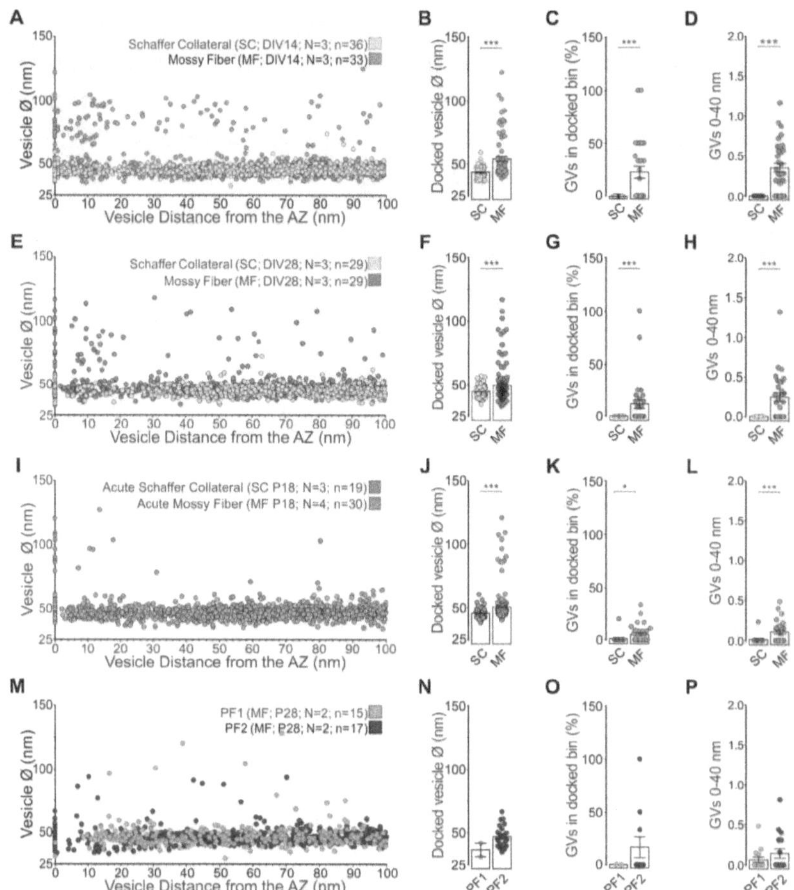

Figure 14. Distribution of giant vesicles in mossy fiber synapses from slice cultures, acute slice preparations, and perfusion-fixed hippocampal tissue.

(A-H) Analysis of Schaffer collateral and mossy fiber synapses from hippocampal slice cultures at DIV14 (A-D) and DIV28 (E-H). (I-L) Analysis of Schaffer collateral and mossy fiber synapses from acute slice preparations from P18 wild-type mice. (M-P) Analysis of mossy fiber synapses from transcardially perfused P28 wild-type mice with two different aldehyde cocktails: PF1 (ice-cold 4% PFA, 2.5% GA in 0.1 M PB), or PF2 (37° C, 2% PFA, 2.5% GA, 2 mM CaCl$_2$ in 0.1 M cacodylate buffer. (A, E, I, M) Clear-core vesicle diameters and their respective distance to the active zone. (B, F, J, N) Scatter plots of docked vesicle diameters. (C, G, K, O) Scatter plots of the respective proportions of the docked vesicle pool occupied by giant vesicles. (D, H, L, P) Scatter plots of the number of giant vesicles within 40 nm of the active zone membrane normalized to active zone area. Statistical significance is represented as *, p<0.05; **, p<0.005; ***, p<0.001. N= number of cultures DIV14 (A-D) and DIV28 (E-H) and number of animals used for acute slice (I-L) and perfusion fixation preparations (M-P); n=number of active zones. See Tables 15-22 for full statistical analyses.

3.2.3. The organization of giant vesicles at mossy fiber active zones is sensitive to aldehyde fixation

In further investigation of *ex vivo* preparations, giant vesicles were observed in mossy fiber-CA3 synapses in wild-type animals perfusion-fixed with either PF1 (ice-cold 4% PFA, 2.5% GA in 0.1 M PB), or PF2 (37° C, 2% PFA, 2.5% GA, 2 mM $CaCl_2$ in 0.1 M cacodylate buffer) (Figure 14 M; Figure 25 A and B). The effects of PF1 were profound; only two docked synaptic vesicles were found in 15 tomograms analyzed (Figure 14 N; 36.67 ± 5.284 nm, PF1; 47.16 ± 1.711 nm, PF2; Figure 25 A and B). Giant vesicles were docked at the active zone membrane in mossy fiber synapses from perfusion fixed animals (Figure 14 O; 0 ± 0, PF1; 16.67 ± 9.796%, PF2), a second *ex vivo* preparation harboring docked giant vesicles. While relatively low in abundance, giant vesicles were present in mossy fiber synapses from both perfusion fixation protocols within 40 nm (Figure 14 P; 0.065 ± 0.035, PF1; 0.145 ± 0.058, PF2; p= 0.45) and 100 nm of the active zone (Table 21; 0.205 ± 0.073, PF1; 0.263 ± 0.076, PF2; p=0.069).

In summary, these data indicate that giant vesicles are not an artifact of slice culture preparations, and this finding is supported by several *ex vivo* preparations. Similar to the consequences of perfusion fixation on synaptic vesicle pools, perfusion fixation depletes docked and membrane-proximal pools of giant vesicles at mossy fiber active zones. However, the potential effects of the perfusion procedure itself (i.e. anoxia) and the lower number of giant vesicles observed within the volume of presynaptic terminals of mossy fibers in *ex vivo* compared to *in vitro* preparations must also be considered when interpreting these results.

3.2.4. Acute pharmacological blockade of network activity does not alter giant vesicle organization

To examine the possibility that spontaneous slice activity contributes to the formation and presynaptic organization of giant vesicles in hippocampal mossy fibers, I treated wild-type slice cultures at DIV14 with drug cocktails designed to pharmacologically silence network activity for 10 minutes prior to HPF. Three conditions were analyzed in this experiment: 1) T/N/A: TTX, to block sodium-propagated action potentials, supplemented with either NBQX and D-AP5 to block AMPA and NMDA receptor blockers, respectively; 2) T/D: TTX with DCG-IV, an mGluR2 receptor agonist that specifically blocks synaptic transmission in mossy fiber synapses (Kamiya and Ozawa, 1999); and 3) VC, comprised of slice culture medium applied in the same manner as the pharmacologically treated slices prior to HPF. Giant vesicles exhibited a similar spatial distribution within 100 nm of mossy fiber active zones after T/N/A and T/D treatment compared to VC (Figure 15 A; Figure 24 A and B). Although the mean diameter of docked vesicles was unchanged after T/N/A and T/D treatment when compared to that of VC mossy fiber synapses (Figure 15 B, 48.84 ± 0.884 nm, VC; 49.29 ± 0.517 nm, T/N/A; 53.86 ± 1.568 nm, T/D; p=0.10), the proportion of giant vesicles comprising the total docked vesicle population was significantly greater after mossy fiber transmission was inhibited by T/D application when compared to T/N/A (Figure 15 C; 6.944 ± 3.943%, VC; 0.926 ± 0.926%, T/N/A; 15.64 ± 4.523%, T/D; p= 0.005). Despite an increase in giant vesicle accumulation within 20 nm of mossy fiber active zones after T/D treatment (Table 25; 0.151 ± 0.047, VC; 0.080 ± 0.031, T/N/A; 0.215 ± 0.063, T/D; p=0.05), the abundance of giant vesicles within 40 nm was not significantly different from VC and T/N/A treatments (Figure 15 D; 0.344 ± 0.080, VC; 0.383 ± 0.063, T/N/A; 0.538 ± 0.088, T/D; p=0.17), indicating giant vesicles are organized closer to the plasma membrane after application of DCG-IV.

These results demonstrate that the relative abundance and spatial distribution of giant vesicles in mossy fiber synapses is largely unaffected by acute pharmacological blockade of network activity in cultured slices. Mossy fiber giant vesicles are therefore unlikely to represent structural endocytic intermediates formed by excessive spontaneous activity in cultured hippocampal slices.

3.2.5. Giant vesicles are present, but do not dock in Munc13-deficient mossy fiber synapses

To more stringently investigate the possibility that mossy fiber giant vesicles represent structural intermediates generated by activity-dependent endocytic mechanisms, I analyzed the spatial organization of giant vesicles in mossy fiber synapses from Munc13-deficient slices. Munc13-deficient synapses exhibit an almost complete loss of spontaneous and evoked synaptic transmission (Augustin et al., 1999; Sigler et al., 2017; Varoqueaux et al., 2002). However, under culture conditions, neurons from Munc13-deficient slices develop normally

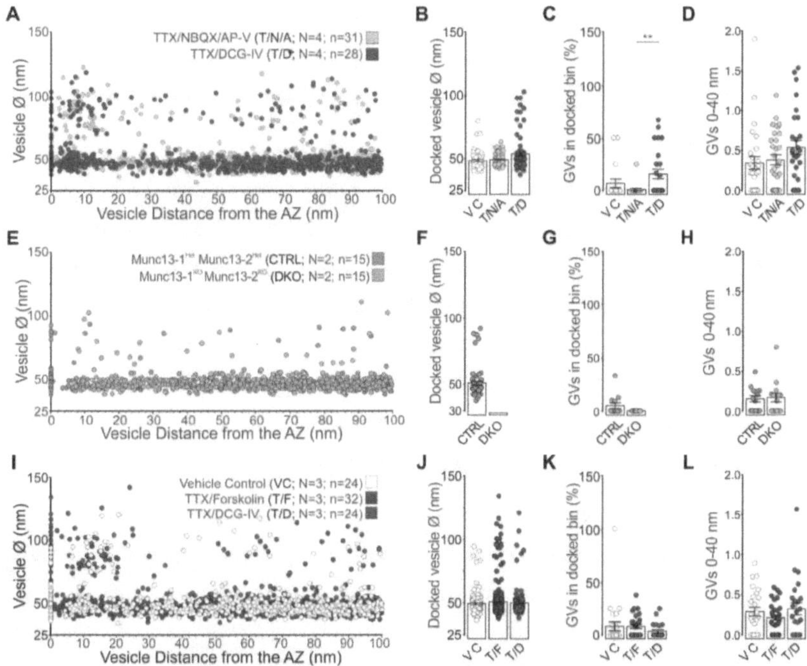

Figure 15. Distribution of giant vesicles in mossy fiber synapses from slice cultures after acute silencing, genetic silencing, or after pharmacological manipulation of release probability.

(**A-D**) Analysis of mossy fiber synapses after a 10 minute pharmacological treatment of slice cultures at DIV14 with one of three conditions: 1) T/N/A: TTX, to block sodium-propagated action potentials, supplemented with either NBQX and D-AP5 to block AMPA and NMDA receptor blockers, respectively; 2) T/D: TTX with DCG-IV, an mGluR2 receptor agonist that specifically blocks synaptic transmission in mossy fiber synapses (Kamiya and Ozawa, 1999); and 3) VC: a vehicle control comprised of slice culture medium. (**E-H**) Analysis of mossy fiber synapses from Munc13-1/2 DKO and CTRL slice cultures at DIV28. (**I-L**) Analysis of mossy fiber synapses after 15 minute treatment with either: 1) T/F: TTX, to block sodium propagated action potentials, and forskolin, an activator of AC1 which increases presynaptic cAMP concentrations (Chavez-Noriega and Stevens, 1994; Dixon and Atwood, 1989; Seamon et al., 1983) thus increasing synaptic transmission (Evans and Morgan, 2003; Seino and Shibasaki, 2005; Weisskopf et al., 1994); 2) T/D: TTX with DCG-IV, and mGluR2 receptor agonist that specifically blocks synaptic transmission in mossy fiber synapses (Kamiya and Ozawa, 1999); and 3) VC: a vehicle control comprised of TTX in slice culture medium. (**A, E, I**) Clear-core vesicle diameters and their respective distance to the active zone membrane. (**B, F, J**) Scatter plots of docked vesicle diameters. (**C, G, K**) Scatter plots of the respective proportions of the docked vesicle pool occupied by giant vesicles. (**D, H, L**) Scatter plots of the number of giant vesicles within 40 nm of the active zone membrane normalized to active zone area. Statistical significance is represented as *, p<0.05; **, p<0.005; ***, p<0.001. N= number of cultures; n=number of active zones. See Tables 23-28 for full statistical analyses.

and establish synaptic connections within anatomically intact circuits. Thus, an ultrastructural examination of mossy fiber synapses in Munc13-deficient slices offers an opportunity to characterize the spatial organization of vesicles in a synapse that has been genetically silenced since "birth" and as such, has never actively participated in synaptic transmission.

Mossy fiber giant vesicles were observed in tomograms of both control and Munc13-1/2 DKO slices (Figure 15 E; Figure 24 C and D). However, no vesicles of any type docked at mossy fiber active zones in Munc13-1/2 DKO slices (Figure 11 E, F, G; 0.051 ± 0.024, CTRL; 0 ± 0, DKO; p=0.10). Instead, vesicles appeared to accumulate at approximately 10 nm from the active zone in mossy fibers from Munc13-1/2 DKO slices (Figure 11 E). Despite this striking docking deficit, the densities of giant vesicles within 40 nm (Figure 15 H; 0.152 ± 0.039, CTRL; 0.169 ± 0.055, DKO; p=0.86) and 100 nm (Table 23; 0.229 ± 0.058, CTRL; 0.315 ± 0.080, DKO; p=0.39) of the active zone membrane were highly comparable between Munc13-1/2 CTRL and DKO mossy fiber synapses.

The results of these experiments effectively exclude the possibility that mossy fiber giant vesicles are formed via activity-dependent forms of compensatory endocytosis (Delvendahl et al., 2016; Watanabe et al., 2013a). My findings demonstrate that like synaptic vesicles, mossy fiber giant vesicles are dependent on Munc13 priming proteins to dock at the active zone membrane. This discovery raises several important questions: Do giant vesicles fuse at active zone release sites? Do giant vesicles contain neurotransmitter and contribute to glutamatergic signaling at mossy fiber synapses?

3.2.6. Mossy fiber giant vesicles are the morphological correlate of giant mEPSCs recorded in CA3 pyramidal neurons

Although my data provide the first unequivocal evidence that giant vesicles dock in physical contact with mossy fiber active zones, their existence has been previously reported (Figure 16 A) (Borges-Merjane et al., 2020; Henze et al., 2002b; Laatsch and Cowan, 1966; Rollenhagen et al., 2007). These observations became of particular importance upon the demonstration that large amplitude (giant) mEPSCs can be recorded from CA3 pyramidal neurons, and that the giant mEPSCs are monoquantal and of mossy fiber origin (Henze et al., 1997, 2002b). Based on the assumption that giant vesicles contain neurotransmitter and are capable of fusing at mossy fiber active zones to generate giant mEPSCs, I rationalized that the relative distribution of mEPSC amplitudes should correlate with the relative proportion of docked giant vesicles visualized and quantified in electron tomograms. To this end, I collaborated with Dr. Bekir Altas, who used patch-clamp electrophysiology to isolate and record mEPSCs from CA3 pyramidal cells in cultured hippocampal slices at DIV14 (Figure 16 B). Initial baselines of all mEPSC events were recorded for a 5-minute period in the presence of TTX, to block sodium-propagated action potentials, and the GABA receptor blocker bicuculline, to exclude contributions from GABAergic transmission (Figure 16 B; black trace). Subsequently, DCG-IV was applied to the bath solution and all DCG-IV-insensitive mEPSCs were recorded for the next 5-minute period (Figure 16 B; gray trace; 5-minute epoch was between the 10th and 15th minute after DCG-IV wash-on). Inhibition of mossy fiber synaptic transmission with DCG-IV reduced the frequency of mEPSC events regardless of amplitude (Figure 16 C). To test whether docked giant vesicles are neurotransmitter-filled, I next compared the frequency of giant mEPSCs to the proportion of docked giant vesicles. To correlate the proportion of giant mEPSC events with the relative proportion of docked giant vesicles at mossy fiber active zones, I subtracted the amplitudes of DCG-IV-insensitive events from the baseline mEPSCs to get an estimate of the DCG-IV-sensitive amplitudes (Figure 16 D, purple line). I then rationalized that the most frequently observed DCG-IV-sensitive mEPSC amplitudes (statistical mode of DCG-IV-sensitive mEPSC amplitudes = 10 pA) reflected fusion and transmitter release from docked vesicles with the most frequently observed dimensions (the statistical mode of docked synaptic vesicle diameters = 44 nm). I calculated the inner lumenal volume of a synaptic vesicle with an outer diameter of 44 nm, accounting for the lipid bilayer (~4 nm of radius measured from tomograms; ~24,400 nm^3 lumenal volume). I then calculated the theoretical mEPSC amplitude that would arise from the fusion of a vesicle at the lower threshold for classification as a giant vesicle (Ø=60 nm; ~ 30 pA; Figure 16 D dotted line). Approximately 27% of DCG-IV-sensitive mEPSCs were larger than 30 pA, in agreement with my finding that approximately 20% of all docked vesicles are giant vesicles in age-matched mossy fiber synapses (Figure 16 C). It is important to note that my calculations are based on several assumptions: i) DCG-IV-sensitive mEPSC events originated from docked

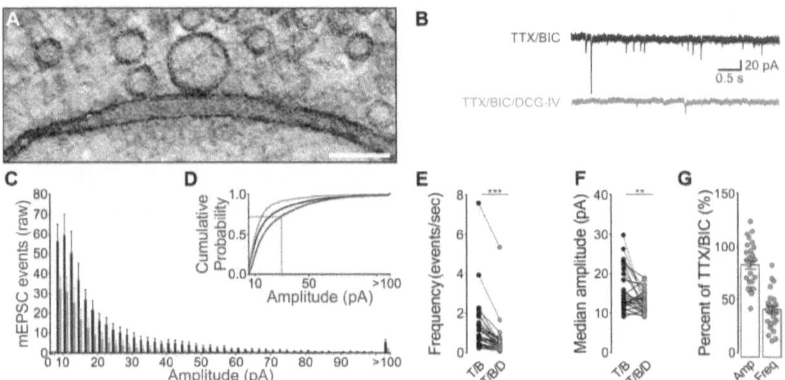

Figure 16. Electrophysiological and morphological analysis of giant vesicles in mossy fiber synapses.

(**A**) Tomographic subvolume of a mossy fiber synapse with a giant vesicle docked directly at the active zone membrane. (**B-G**) Effects of DCG-IV on mEPSC events recorded in CA3 pyramidal neurons in hippocampal slice cultures at DIV14. (**B**) Example traces of mEPSCs recorded from CA3 pyramidal neurons in the presence of 1 µM TTX and 10 µM bicuculline (BIC) before (black trace) and after the wash on of 2 µM DCG-IV (grey trace). (**C**) Frequency distribution of mEPSC amplitudes recorded in CA3 pyramidal neurons before (black) and after (gray) application of DCG-IV (2 pA bins from 8 pA until 100 pA and all events greater than 100 pA are pooled in one bin). (**D**) Cumulative distribution of mEPSC amplitude before and after application of DCG-IV. (**E**) Before-after scatter plot of mEPSC frequency before (T/B) and after (T/B/D) application DCG-IV. (**F**) Before-after scatter plot of median mEPSC amplitude before and after the application of DCG-IV. (**G**) Relative changes in the median amplitude and frequency after the application of DCG-IV normalized to TTX/BIC. Statistical significance is represented as *, p<0.05; **, p<0.005; ***, p<0.001. N=2 cultures; n=28 cells. Scale bar: 100 nm, **A**. See Table 29 for full statistical analysis.

synaptic vesicles at mossy fiber synapses; ii) synaptic vesicle filling was proportional to the size of a given vesicle (Bruns et al., 2000); and iii) postsynaptic receptor saturation was negligible.

The amplitudes of the remaining DCG-IV-insensitive mEPSCs were reduced compared to amplitudes of all mEPSCs (Figure 16 D). DCG-IV caused a significant reduction in the mEPSC frequency in CA3 pyramidal neurons compared to all mEPSC events before the application of DCG-IV (Figure 16 E; 1.376 ± 0.273 events/sec, TTX/BIC; 0.60 ± 0.186 events/sec, TTX/BIC/DCG-IV; p<0.001). The median amplitude of mEPSC events recorded in CA3 pyramidal neurons was significantly reduced after DCG-IV application (Figure 16 F; 16.25 ± 1.031 pA, TTX/BIC; 13.0 ± 0.524 pA, TTX/BIC/DCG-IV; p=0.007). The specific inhibition of mossy fiber synaptic transmission with DCG-IV demonstrates that 41% of mEPSCs onto CA3 pyramidal neurons are insensitive to DCG-IV (Figure 16 G; 40.75 ± 3.20%), and likely arise from excitatory collaterals from other CA3 pyramidal neurons described to form in rat slice cultures (Frotscher and Gähwiler, 1988). Furthermore, the median amplitude of DCG-IV-insensitive events was reduced to about 82% of all mEPSC events measured prior to the application of DCG-IV (Figure 16 G; 82.43 ± 3.939%), meaning that many but not all giant mEPSC events were sensitive to DCG-IV. It is unclear the extent at which DCG-IV inhibits spontaneous vesicle fusion at hippocampal mossy fiber synapses however these results are in agreement with the reduction in mEPSC events after gamma-irradiation of hippocampal granule cells (Henze et al., 1997).

My findings indicate a close correlation between the relative proportions of DCG-IV-sensitive giant mEPSCs and docked giant vesicles at mossy fiber synapses, thereby supporting the hypothesis that giant vesicles contribute to glutamatergic signaling at mossy fiber-CA3 synapses. A potential contribution of multivesicular release events to the larger amplitude mEPSCs recorded in CA3 pyramidal neurons cannot be completely excluded. My data support the notion that giant vesicles indeed contain neurotransmitter and that they have the potential to profoundly influence synaptic transmission at mossy fiber-CA3 synapses (Henze et al., 2002b). I therefore investigated whether changes in synaptic strength correlate with corresponding changes in the numbers of docked, and presumably fusion-competent, giant vesicles.

3.2.7. Pharmacological manipulation of presynaptic cAMP does not alter giant vesicle organization

If giant vesicles represent bona fide synaptic vesicles with the capacity to release large quanta of neurotransmitter, I rationalized that alterations in their availability at active zone release sites may contribute to changes in mossy fiber synaptic transmission efficacy upon manipulation of presynaptic cAMP levels. I treated wild-type slice cultures at DIV28 with drug cocktails to either increase or decrease presynaptic cAMP concentrations. Three conditions were analyzed in this experiment: 1) T/F: TTX, to block sodium propagated action potentials, and forskolin, an activator of AC1 which increases presynaptic cAMP concentrations (Chavez-Noriega and Stevens, 1994; Dixon and Atwood, 1989; Seamon et al., 1983) thus increasing synaptic transmission (Evans and Morgan, 2003; Seino and Shibasaki, 2005; Weisskopf et al., 1994); 2) T/D: TTX with DCG-IV, an mGluR2 receptor agonist that specifically blocks synaptic transmission in mossy fiber synapses (Kamiya and Ozawa, 1999); and 3) VC: comprised of TTX in slice culture medium. Consistent with data obtained from untreated mossy fiber synapses at DIV28, mossy fiber synapses harbored giant vesicles distributed throughout the synaptic subvolume following pharmacological manipulation of presynaptic cAMP levels (Figure 15 I; Figure 24 E and F). Giant vesicles docked at mossy fiber synapses in VC, T/F, and T/D (Figure 15 J; 49.26 ± 0.631, VC; 50.69 ± 0.662, T/F; 49.68 ± 0.754, T/D; p=0.94), however the proportion of docked giant vesicles per active zone was slightly lower when presynaptic levels of cAMP were reduced with DCG-IV, although this tendency did not reach statistical significance (Figure 15 K; 8.241 ± 4.179%, VC; 8.012 ± 1.677%, T/F; 3.504 ± 1.420%, T/D; p=0.09). The densities of giant vesicles were comparable regardless of presynaptic cAMP concentrations within 40 nm (Figure 15 L; 0.288 ± 0.055, VC; 0.215 ± 0.030, T/F; 0.325 ± 0.075, T/D; p=0.62) and within 100 nm of the active zone (Table 27; 0.561 ± 0.098, VC; 0.333 ± 0.44, T/F; 0.476 ± 0.093, T/D; p=0.93).

These data demonstrate that the organization of giant vesicles at mossy fiber active zones is insensitive to acute, pharmacologically induced changes in presynaptic cAMP concentrations. Therefore, forskolin-induced enhancement of mossy fiber transmitter release efficacy is unlikely to derive from increased availability or fusion of giant vesicles at mossy fiber active zones.

3.2.8. Giant vesicles may originate in granule cell somas and are not restricted to immature mossy fiber synapses

Although the existence of giant vesicles in mossy fiber synapses is well-documented (Henze et al., 2002b; Laatsch and Cowan, 1966; Rollenhagen et al., 2007), the origin of this vesicle type in mossy fiber synapses is unknown. While it is possible that giant vesicles are precursor vesicles that traffic from granule cell somata, the molecular identity of giant vesicles has yet to be characterized. To this end, I took low magnification electron tomograms of mossy fiber axonal projections from granule cells in acute hippocampal slices. All vesicle types docked at mossy fiber active zones were observed in electron tomograms of mossy fiber axonal projections from granule cells to the CA3 in acute hippocampal slices from P18 wild-type mice (Figure 17 A-I). These vesicles include synaptic vesicles (Figure 17 A; open arrowheads), giant vesicles (Figure 17 A, B, E, F, G, H, I; grey arrowheads), and DCVs (Figure 17 A, C, D, H). Giant vesicles in the mossy fiber axons are positioned within nanometers of axonal microtubules

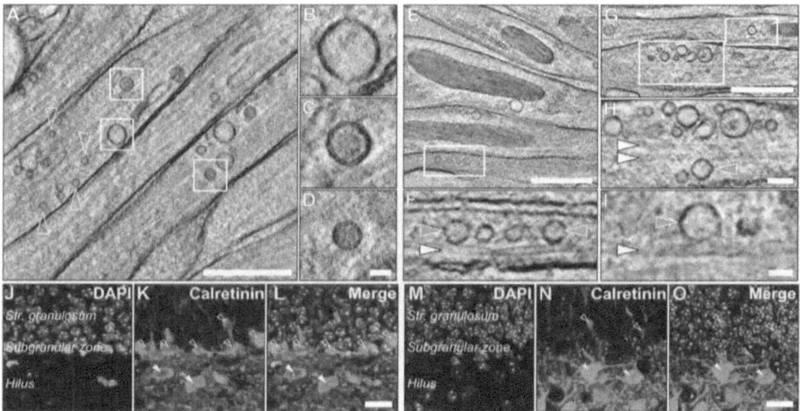

Figure 17. Giant vesicles may have somatic origins and are not merely an early developmental phenomenon.

(A-I) Tomographic subvolume of low-magnification electron tomograms taken of mossy fiber axonal projections from acute slice preparations of P18 wild-type mice. All three vesicle types that are observed docked at mossy fiber active zones, synaptic vesicles (open arrowheads), giant vesicles (B), and DCVs (C-D) Positioning of synaptic vesicles, giant vesicles, and DCVs in proximity to axonal microtubules (E-I) indicate trafficking of these vesicles within the granule cell axons. Giant vesicles indicated with gray arrowheads, microtubules with white arrowheads (F, H-I). (J-O) Confocal images of the subgranular zone in the dentate gyrus from acute hippocampal slices (J-L) and slice cultures at DIV28 (M-O) where calretinin-expressing (K and N; green) new born granule cells are observed in the acute slice (K; open arrowheads), but rarely in the slice culture (N). (J and M; visualization of nuclei with DAPI, cyan). Calretinin-positive hilar mossy cells (K-L; N-O; white arrowheads) are unchanged in slice cultures (N-O) compared to acute slices (K-L). (L, O) Merged confocal images of DAPI and calretinin labeling in acute slices (L) and slice culture (O). Scale Bars: 500 nm, A, E, G; 100 nm, H; 50 nm, B-D, P 20 µm, J-O.

(Figure 17 F, H, I; white arrowheads), indicative of anterograde trafficking to the synaptic bouton. This evidence, however, arises from images of fixed tissues. Additional experimentation is necessary to determine the direction of the axonal giant vesicles.

Neurogenesis in the hippocampal DG continues well into adulthood in rodents (Mongiat and Schinder, 2011). To exclude the possibility that giant vesicles were present in immature mossy fiber boutons and thus an artifact of newly generated granule cells, I compared the number of newly generated granule neurons from ex vivo preparations and age-matched hippocampal slice cultures. Newly generated calretinin-positive granule cells (Brandt et al., 2003) found in the subgranular zone of the DG in acute brain slices (open arrowheads; Figure 17 J-L) were almost completely absent in hippocampal slice cultures (Figure 17M-O), indicating an almost complete cessation of granule cell neurogenesis in our model system. In cultured hippocampal tissue, calretinin immunoreactivity was restricted to hilar mossy cells (white arrowheads in Figure 17 K, L, N, O). These findings indicate that the analysis of mossy fiber synapses from hippocampal slice cultures did not unintentionally include immature mossy fiber boutons formed by newborn granule cells during the culture period, therefore demonstrating that giant vesicles are not an artifact of mossy fiber development. These results do not exclude the possibility that other mechanisms of giant vesicle formation may exist and indicate that further investigation is required.

3.3. DCVs dock at the active zone in mossy fiber synapses

DCVs are vesicular organelles that transport and secrete peptide signaling molecules in many cells within a living organism. In the central nervous system, DCVs package, transport, and release neuropeptides that modulate the function of synaptic transmission on the pre- and postsynaptic cell (Salio et al., 2006). DCVs can be induced to fuse at both synaptic and extrasynaptic sites (van de Bospoort et al., 2012; Farina et al., 2015; Tao et al., 2018a), however, ultrastructural analyses have typically detected them at a distance from the active zone (van de Bospoort et al., 2012; Cifuentes et al., 2008; Imig et al., 2014). Consistent with the relatively high abundance of neuropeptides produced by dentate granule cells (Danzer and McNamara, 2004; Derrick et al., 1992; Salin et al., 1995; Simmons and Chavkin, 1996; Weisskopf et al., 1993), and with previous observations of DCVs within mossy fiber synapses in *ex vivo* preparations (Chicurel and Harris, 1992; Commons and Milner, 1996; Dieni et al., 2012, 2015; Rollenhagen et al., 2007; Sadakata et al., 2013), my data demonstrate that DCVs not only accumulate, but actually dock in physical contact with the active zone membrane.

3.3.1. DCVs dock at mossy fiber active zones in slice cultures and acute slice preparations

In mossy fiber synapses from hippocampal slice cultures at both DIV14 and DIV28, DCVs were observed within 100 nm of the active zone and had a tendency to accumulate within 20 nm of the active zone (Figure 18 A; DIV14 green, DIV28 grey). Similar numbers of DCVs docked at mossy fiber active zones at both developmental time points (Figure 18 B; 0.075 ± 0.023, DIV14; 0.050 ± 0.019, DIV28). The majority of DCVs measured were located within 40 nm of the active zone (Figure 18 C, 0.163 ± 0.036, DIV14; 0.190 ± 0.032, DIV28; compared to Figure 18 D, 100 nm, 0.278 ± 0.049, DIV14; 0.282 ± 0.047, DIV28). In contrast, no DCVs were found within 100 nm of the active zone in Schaffer collateral synapses at either DIV14 or DIV28, although a number of DCVs were observed beyond the 100 nm cutoff (data not shown).

In acute slice preparations from P18 wild-type mice, DCVs were also found within 100 nm of the active zone membrane in both mossy fiber and Schaffer collateral synapses (Figure 18 E, Schaffer collateral dark grey, mossy fiber green). Mossy fiber synapses in acute slices also harbored a comparable number of docked DCVs to age-matched slice cultures (Figure 18 F

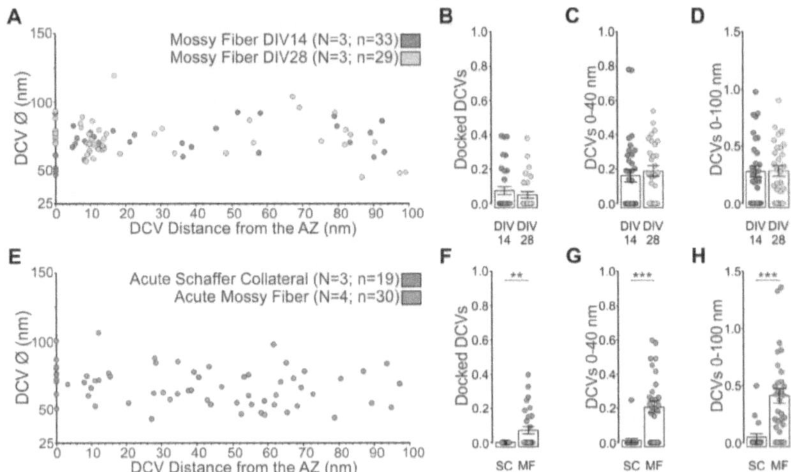

Figure 18. Spatial distribution of DCVs in mossy fiber synapses in slice cultures and acute slice preparations.

(A-D) Analysis of DCV distribution in mossy fiber synapses from hippocampal slice cultures at DIV14 (green) and DIV28 (gray). (E-H) Analysis of DCV distribution in Schaffer collateral and mossy fiber synapses from acute slice preparations from P18 wild-type mice. (A, E) DCV diameters and their respective distance to the active zone membrane. (B, F) Scatter plot of the number of docked DCVs normalized to active zone area. (C, D, G, H) Scatter plots of the number of DCVs within 40 nm (C, G) and 100 nm (D, H) of the active zone membrane. Statistical significance is represented as *, p<0.05; **, p<0.005; ***, p<0.001. N=number of cultures (A-D), number of animals (E-H); n=number of active zones. See Tables 15-20 for full statistical analyses.

acute slice, 0.072 ± 0.021, MF; compared to DIV14 slice culture, Figure 18 B). In the acute slice preparation, DCVs were evenly distributed within 100 nm of the active zone membrane and did not cluster within 20 nm of the active zone membrane (Figure 18 E). The number of DCVs within 40 nm of the active zone membrane was comparable between mossy fiber synapses in acute slices and in age-matched slice cultures (Figure 18 G, 0.013 ± 0.013, SC; 0.209 ± 0.034, MF; p<0.001; compared to DIV14, Figure 18 C). However, the density of DCVs within 100 nm was higher in the acute slice preparation than in mossy fiber synapses from hippocampal slice cultures at DIV14 (Figure 18 H; 0.048 ± 0.029, SC; 0.411 ± 0.065, MF; p<0.001). In 19 Schaffer collateral synapses, there were five DCVs measured within 100 nm of the active zone; the majority of which were located more than 40 nm from the active zone membrane (Figure 18 E, G, H).

My findings indicate that DCV docking at hippocampal mossy fiber synapses also occurs *in vivo* and therefore does not represent an artifact introduced by the slice culture procedure. Moreover, my comparative analysis indicates that DCV docking directly at the active zone membrane is specific to, or occurs considerably more frequently at, mossy fiber synapses.

3.3.2. Munc13 priming proteins are essential for DCV docking at mossy fiber synapses and lead to accumulation of DCVs in proximity to the active zone

Genetically silenced Munc13-1/2 DKO mossy fiber synapses exhibited a complete loss of docked synaptic vesicles (Figure 11 F), giant vesicles (Figure 15 G), and DCVs (Figure 19 B). Moreover, DCVs in mossy fiber synapses from Munc13-1/2 DKO slice cultures accumulated between 10 and 20 nm from the active zone (Figure 19 A), as previously observed for synaptic vesicles (Figure 11 F) and giant vesicles (Figure 15 E). The spatial density of docked DCVs was comparatively lower in Munc13-1/2 CTRL slices compared to wild-type slices of a similar age (Figure 19 B; 0.077 ± 0.035, CTRL; 0 ± 0, MF; p=0.10), presumably due to the relatively small number of tomograms analyzed in this experiment. Despite the lack of docked DCVs in Munc13-1/2 DKO mossy fiber synapses, the density of DCVs within 40 nm of the active zone was comparable with that of Munc13-1/2 CTRL littermates (Figure 19 C; 0.191 ± 0.054, CTRL; 0.343 ± 0.072, DKO; p=0.15). Unexpectedly, Munc13-1/2 DKO mossy fiber synapses had a higher accumulation of DCVs within 100 nm of the active zone compared to CTRL littermates (Figure 19 D; 0.252 ± 0.061, CTRL; 0.643 ± 0.095, DKO; p=0.002). The accumulation of DCVs in Munc13-1/2 DKO mossy fiber synapses was the highest measured in mossy fiber synapses from hippocampal slice cultures.

These results indicate that Munc13 priming proteins are essential for DCV docking at hippocampal mossy fiber active zones. Moreover, the membrane-proximal accumulation of DCVs in Munc13-deficient mossy fiber synapses indicates that DCVs may undergo a tethering step prior to docking and priming at active zones. Finally, increased abundance of DCVs within 100 nm of the active zone membrane in Munc13-deficient mossy fiber synapses provides indirect evidence of continued anterograde transport of DCVs from the soma even in the absence of synaptic transmission. Whether increased abundance reflects the loss of basal DCV

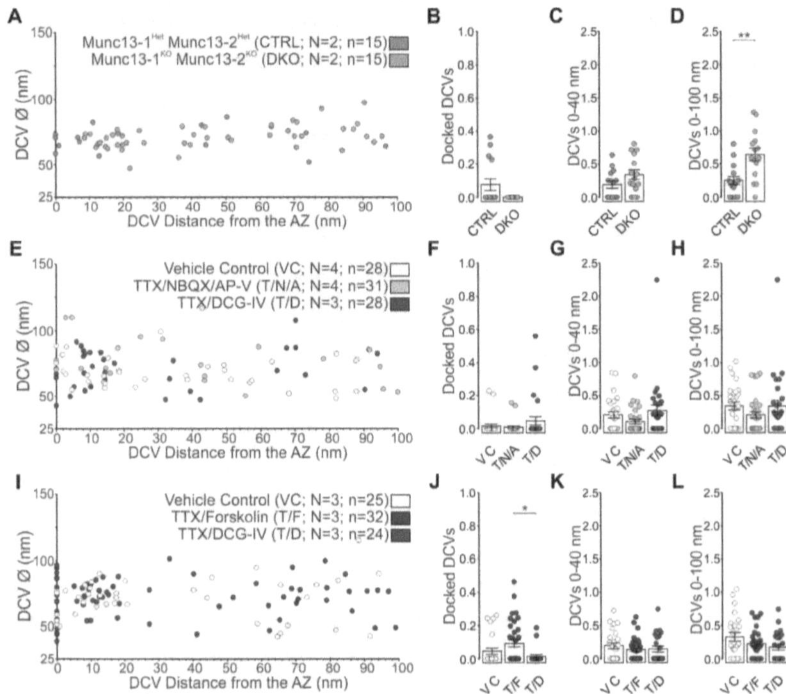

Figure 19. Spatial distribution of DCVs in mossy fiber synapses in genetically and pharmacologically manipulated slice cultures.

(**A-D**) Analysis of DCV distribution in mossy fiber synapses from Munc13-1/2 DKO and CTRL slice cultures at DIV28. (**E-H**) Analysis of DCV distribution in mossy fiber synapses after a 10 minute pharmacological silencing of slice cultures at DIV14 with one of three conditions: 1) T/N/A: TTX, to block sodium-propagated action potentials, supplemented with either NBQX and D-AP5 to block AMPA and NMDA receptor blockers, respectively; 2) T/D: TTX with DCG-IV, an mGluR2 receptor agonist that specifically blocks synaptic transmission in mossy fiber synapses (Kamiya and Ozawa, 1999); and 3) VC: comprised of slice culture medium. (**I-L**) Analysis of DCV distribution in mossy fiber synapses after a 15 minute treatment with either: 1) T/F: TTX, to block sodium propagated action potentials, and forskolin, an activator of AC1 which increases presynaptic cAMP concentrations (Chavez-Noriega and Stevens, 1994; Dixon and Atwood, 1989; Seamon et al., 1983) thus increasing synaptic transmission (Evans and Morgan, 2003; Seino and Shibasaki, 2005; Weisskopf et al., 1994); 2) T/D: TTX with DCG-IV, and mGluR2 receptor agonist that specifically blocks synaptic transmission in mossy fiber synapses (Kamiya and Ozawa, 1999); and 3) VC: comprised of TTX in slice culture medium in hippocampal slice cultures at DIV28. (**A, E, I**) Diameters of all DCVs analyzed and their respective distance to the active zone membrane. (**B, F, J**) Scatter plot of the number of docked DCVs normalize to active zone area. (**C, D, G, H, K, L**) Scatter plots of the number of DCVs within 40 nm (**C, G, K**) and 100 nm (**D, H, L**) of the active zone membrane normalized to active zone area. Statistical significance is represented as *, p<0.05; **, p<0.005; ***, p<0.001. N=number of cultures; n=number of active zones. See Tables 23-28 for full statistical analyses.

fusion in Munc13-1/2 DKO mossy fiber terminals, or a compensatory increase in the production and trafficking of DCVs to mossy fiber terminals remains to be determined.

3.3.3. Acute pharmacological silencing of mossy fiber synapses does not change the accumulation and distribution of DCVs

To determine whether DCV docking was influenced by spontaneous network activity, wild-type hippocampal slice cultures at DIV14 were pharmacologically silenced shortly before cryofixation. Three conditions were analyzed in this experiment: 1) T/N/A: medium supplemented with TTX, to block sodium-propagated action potentials, and NBQX and D-AP5 to block AMPA and NMDA receptor blockers, respectively; 2) T/D: medium supplemented with TTX and DCG-IV, an mGluR2 receptor agonist that specifically blocks synaptic transmission in mossy fiber synapses (Kamiya and Ozawa, 1999); and 3) VC, comprised of slice culture medium applied in the same manner as the pharmacologically treated slices prior to HPF. The spatial distribution of DCVs in mossy fiber synapses were comparable across the three treatment conditions (Figure 19 E). The spatial density of docked DCVs at mossy fiber active zones was unchanged between pharmacologically silenced and control slices (Figure 19 F; 0.014 ± 0.010, VC; 0.020 ± 0.013, T/N/A; 0.043 ± 0.023, T/D; p=0.63). The abundance of DCVs within 40 nm (Figure 19 G; 0.260 ± 0.048, VC; 0.104 ± 0.034, T/N/A; 0.271 ± 0.082, T/D; p=0.05) or within 100 nm of mossy fiber active zones (Figure 19 H; 0.342 ± 0.060, VC; 0.205 ± 0.049, T/N/A; 0.337 ± 0.087, T/D; p= 0.17) was comparable between pharmacologically silenced and control slices.

These findings indicate that acute pharmacological blockade of spontaneous network activity does not substantially alter the spatial distribution of DCVs in mossy fiber synapses. Classically, synaptic DCV fusion occurs during states of persistent, high activity (i.e. multiple trains of high frequency stimulation; van de Bospoort et al., 2012; Farina et al., 2015) and it is likely spontaneous network activity does not involve neuropeptide signaling in the hippocampal mossy fiber-CA3 synapse.

3.3.4. Pharmacological manipulation of presynaptic cAMP affects DCV distribution and docking in mossy fiber synapses

To determine whether presynaptic cAMP had an effect on DCV organization in mossy fiber synapses, I treated wild-type slice cultures at DIV28 with drug cocktails to either increase or decrease presynaptic cAMP. Three conditions were analyzed in this experiment: 1) T/F: TTX, to block sodium propagated action potentials, and forskolin, an activator of AC1 which increases the production of presynaptic cAMP (Chavez-Noriega and Stevens, 1994; Dixon and Atwood, 1989; Seamon et al., 1983) thus increasing synaptic transmission (Evans and Morgan, 2003; Seino and Shibasaki, 2005; Weisskopf et al., 1994); 2) T/D: TTX with DCG-IV, an mGluR2 receptor agonist that specifically blocks synaptic transmission in mossy fiber synapses (Kamiya and Ozawa, 1999); and 3) VC: comprised of TTX in slice culture medium. DCVs were observed in mossy fiber synapses after pharmacological manipulations of presynaptic cAMP, as well as in VC conditions (Figure 19 I). Forskolin-induced increases in presynaptic cAMP caused an increase in the spatial density of docked DCVs and a decrease in DCV docking upon reduction of presynaptic cAMP with DCG-IV (Figure 19 J; 0.045 ± 0.019, VC; 0.092 ± 0.022, T/F; 0.014 ± 0.009, T/D; p=0.014). These findings indicate that DCV docking is regulated by cAMP-dependent mechanisms operating very close to the active zone, since the density of membrane-proximal DCVs within 40 nm of the active zone (Figure 19 K; 0.192 ± 0.045, VC; 0.133 ± 0.028, T/F; 0.141 ± 0.043, T/D; p=0.61) or within 100 nm of the active zone (Figure 19 L; 0.328 ± 0.064, VC; 0.225 ± 0.036, T/F; 0.170 ± 0.044, T/D; p=0.19) was not significantly different between forskolin- or DCG-IV-treated slice cultures.

These findings indicate that acute pharmacological manipulations known to induce changes in mossy fiber synaptic transmission via manipulation of presynaptic cAMP concentrations result in corresponding changes in DCV docking at active zone membranes. It is tempting to speculate that forskolin-induced increases in DCV docking contribute to cAMP-mediated increases in release probability, since neuropeptides have been implicated in the modulation of mossy fiber transmission (Henze et al., 2000; Salin et al., 1995; Weisskopf et al., 1993). However, alternative possibilities, including the DCV-mediated delivery of active zone components during *de novo* active zone formation (Sorra et al., 2006; Tao et al., 2018a) must also be considered.

3.4. Morphological RRP estimates from past and present studies

This is the first study to systematically investigate presynaptic mossy fiber ultrastructure and the organization of vesicles at active zone release sites using electron tomography. In this section, I extrapolated the spatial density and dimensions of docked vesicles reconstructed in tomographic subvolumes to the level of individual active zones and to the level of entire presynaptic mossy fiber terminals. Using these estimates, I compared my data with previous morphological (Rollenhagen et al., 2007) and functional (Hallermann et al., 2003; Midorikawa and Sakaba, 2017) estimates of RRP size in hippocampal mossy fiber synapses.

To compare the present study to past studies, I first normalized my data by multiplying the relative spatial density of docked synaptic vesicles, giant vesicles, and DCVs, respectively, to the average active zone area reported by Rollenhagen and colleagues (mean active zone area = 0.12 μm^2 determined by 3D serial section EM; Rollenhagen et al., 2007). In mossy fiber synapses at DIV14, I calculated approximately 5.5 synaptic vesicles, 1.4 giant vesicles, and 0.9 DCVs docked per active zone (Figure 11). At DIV28, approximately 11.6 synaptic vesicles, 1.1 giant vesicles, and 0.6 DCVs docked per active zone (Figure 12). The total number of active zones per mossy fiber bouton determined by 3D serial section EM reconstructions of entire terminals from past studies ranges from 18-45 (as many as 37 active zones reported by Chicurel and Harris, 1992; as many as 45 active zones and average of 29.75 active zones reported by Rollenhagen et al., 2007; an average of 25.3 active zones reported by Sai et al., 2017). Assuming an average of 29.75 active zones per mossy fiber bouton, approximately 163 synaptic vesicles, 42 giant vesicles, and 27 DCVs would be morphologically docked per mossy fiber bouton at DIV14. At DIV28, approximately 345 synaptic vesicles, 33 giant vesicles, and

Table 11. Estimating the total change in membrane capacitance of docked and membrane-proximal vesicles in cultured mossy fiber synapses at DIV14.

MF Synapses at DIV14	Vesicle Type	Mean Diameter (nm)	Mean #/AZ	Surface Area/AZ (nm^2)	Estimated ΔC_m/ Bouton (fF)
Docked Vesicles	SV (Ø<60 nm)	46.8	5.5	37785	
	GV (Ø>60 nm)	85.5	1.4	32558	
	DCV	68.4	0.9	13244	
	Total			83587	24-37
Vesicles 0-40 nm	SV (Ø<60 nm)		20.6	132322	
	GV (Ø>60 nm)		4.3	98522	
	DCV		2.0	34014	
	Total			264858	78-119

Abbreviations: AZ, active zone; ΔC_m, change in membrane capacitance; GV, giant vesicle; MF, mossy fiber; SV, synaptic vesicle.

18 DCVs would be docked per mossy fiber bouton. Based on the mean diameter of respective vesicle types obtained from tomographic reconstructions, I calculated the respective vesicular surface areas that would be integrated with the plasma membrane upon fusion of all docked vesicles (Table 11, DIV14; Table 12, DIV28). Based on a specific membrane capacitance of 1 μF/cm^2 (Gentet et al., 2000; Hallermann et al., 2003; Midorikawa and Sakaba, 2017), my data would predict membrane capacitance increases of between 24-37 fF at DIV14 (Table 11) and 33-50 fF at DIV28 (Table 12) if all morphologically docked vesicles were to fuse during a step depolarization. These estimates appear compatible with the membrane capacitance increase reported to correspond with depletion of the RRP in cultured mossy fiber terminals (Table 13; Midorikawa and Sakaba, 2017).

Table 12. Estimating the total change in membrane capacitance of docked and membrane-proximal vesicles in cultured mossy fiber synapses at DIV28.

MF Synapses at DIV28	Vesicle Type	Mean Diameter (nm)	Mean #/AZ	Surface Area/AZ (nm^2)	Estimated Change in Capacitance/ Bouton (fF)
Docked Vesicles	SV ($\varnothing$<60 nm)	45.7	11.6	76136	
	GV ($\varnothing$>60 nm)	84.3	1.1	24898	
	DCV	78.0	0.6	11471	
	Total			112505	33-50
Vesicles 0-40 nm	SV ($\varnothing$<60 nm)		25.1	161281	
	GV ($\varnothing$>60 nm)		3.0	68827	
	DCV		2.3	39649	
	Total			269757	80-121

Abbreviations: AZ, active zone; ΔC_m, change in membrane capacitance; GV, giant vesicle; MF, mossy fiber; SV, synaptic vesicle.

Based on the assumption that membrane-proximal vesicles positioned upstream of docking and priming might be recruited to the functional RRP and elicited to fuse during strong depolarizing pulses, I calculated the theoretical change in membrane capacitance if all vesicles within 0-40 nm of the active zone membrane were to fuse with the presynaptic membrane at DIV14 (Table 11) and DIV28 (Table 12). The theoretical change in membrane capacitance at DIV28 would range between 80 and 121 fF, very close to the larger change in presynaptic

Table 13. Published presynaptic capacitance studies of mossy fiber boutons.

Method	Publication	Preparation	Pulse Duration	ΔC_m/Bouton (fF)
Electrophysiology	(Hallermann et al., 2003)	Rat, acute slice	30 ms	100
	(Midorikawa and Sakaba, 2017)	Rat, acute slice	30 ms	50
	(Midorikawa and Sakaba, 2017)	Rat, dissociated	30 ms	30

Abbreviations: ΔC_m, change in membrane capacitance

membrane capacitance recorded by Hallermann and colleagues (Table 13; Hallermann et al., 2003).

These findings indicate that the number of morphologically docked vesicles correlates relatively well with functional estimates of the RRP assessed by presynaptic capacitance recordings in hippocampal mossy fibers. However, it indicates that strong stimuli applied by direct depolarization of the presynaptic terminal likely trigger fusion of more than just the membrane-attached pool of docked vesicles. It should however be noted that hippocampal mossy fibers exhibit a very low initial release probability (Figure 8) (Lawrence et al., 2004; Nicoll and Schmitz, 2005; Salin et al., 1996; Vyleta and Jonas, 2014), and that action potentials likely evoke fusion from only a subpopulation of the docked vesicle pool. It will therefore be interesting to investigate potential heterogeneity in the vesicular release probability of different vesicle classes in future studies. Nevertheless, my work represents the first study to take the morphological heterogeneity of presynaptic vesicle pools into account in a systematic ultrastructural analysis of mossy fiber active zone organization.

4. Discussion

4.1. Overview

Synaptic transmission operates by a stereotypical principle involving multiple molecularly regulated steps prior to vesicle fusion and transmitter release at presynaptic active zones. This complexity provides multiple opportunities for the modulation of synaptic transmitter release efficacy. Several mechanisms have been postulated to play a role in determining synaptic release probability, including the physical distance between calcium influx and calcium sensor (Chen et al., 2015), the type of calcium buffer present (Müller et al., 2005; Vyleta and Jonas, 2014), proximity of mitochondria to active zone release sites (Brodin et al., 1999; Smith et al., 2016), the intrinsic properties of synaptic vesicles related to the state of the release machinery (Cano et al., 2012), and the availability of functionally primed and release-competent vesicles (Imig et al., 2014). Based on several converging lines of experimentation that indicate considerable overlap between pools of morphologically docked and functionally primed vesicles (Imig et al., 2014; Rosenmund and Stevens, 1996; Schikorski and Stevens, 1997, 2001; Siksou et al., 2009a), I performed experiments to address the question of whether the availability of morphologically docked synaptic vesicles contributes to differences in the initial synaptic release probability of hippocampal Schaffer collateral and mossy fiber synapses. I adopted a methodological approach combining hippocampal organotypic slice culture, HPF, AFS, and electron tomography to accurately resolve synaptic ultrastructure in a near-native and 3D context. In my comparative ultrastructural analysis of presynaptic vesicle organization at wild-type hippocampal Schaffer collateral and mossy fiber synapses at rest, I discovered that low release probability mossy fiber synapses are characterized by a low spatial density of docked vesicles at individual active zones. Mossy fiber synapses were additionally distinguished by the presence of a prominent membrane-proximal and potentially tethered pool of vesicles. I hypothesize that this membrane-proximal pool is ideally situated to rapidly resupply the pool of docked and primed vesicles during sustained synaptic activity and that it likely contributes to the characteristic facilitation exhibited by mossy fiber synapses (Marchal and Mulle, 2004; Nicoll and Schmitz, 2005; Salin et al., 1996).

My systematic ultrastructural analysis revealed that three distinct types of vesicle dock at mossy fiber active zones, including synaptic vesicles, giant vesicles, and DCVs. Moreover, I demonstrate that this morphological heterogeneity within the docked vesicle pool is also present at mossy fiber active zones *in vivo* and therefore not an artifact of the slice culture preparation. Although the origin and functional relevance of giant vesicles remains to be determined, my data support the hypothesis that giant vesicles contain neurotransmitter and contribute to glutamatergic signaling at mossy fiber active zones. By extending my analyses to include genetically silenced synapses, I demonstrated that similar Munc13-dependent molecular mechanisms operate to dock and functionally prime all three species of vesicle at mossy fiber synapses.

Consistent with a significant overlap between pools of morphologically docked and functionally primed vesicles (Imig et al., 2014; Rizzoli and Betz, 2004; Rosenmund and Stevens, 1996; Schikorski and Stevens, 1997, 2001; Siksou et al., 2009a), my data indicate a relatively close correlation between numbers of membrane attached vesicles and functional RRP estimates estimated from presynaptic capacitance recordings (Hallermann et al., 2003; Midorikawa and Sakaba, 2017). Importantly, my data indicate that protocols applying strong stimulation, such as direct depolarization of the presynaptic terminal, likely trigger the fusion of both membrane-proximal and docked vesicles. Since postsynaptic responses elicited by presynaptic action potentials appear to result from the fusion of only a small subpopulation of the total docked pool (Bekkers and Stevens, 1995; Gustafsson et al., 2019; Hanse and Gustafsson, 2001; Stevens and Tsujimoto, 1995; Vyleta and Jonas, 2014), my data indicate heterogeneity in the vesicular release probability of membrane-attached vesicles docked at mossy fiber active zones.

In addition to their morphological characteristics, hippocampal mossy fiber synapses are functionally distinguished by their almost exclusive reliance on presynaptically expressed plasticity mechanisms (Lawrence et al., 2004; Maccaferri et al., 1998; Nicoll and Schmitz, 2005; Salin et al., 1996; Toth et al., 2000). I investigated whether pharmacological manipulation of presynaptic cAMP (Huang et al., 1994b; Seino and Shibasaki, 2005; Tzounopoulos et al., 1998; Villacres et al., 1998; Weisskopf et al., 1994) caused corresponding changes in presynaptic vesicle organization at mossy fiber active zones. My analysis of forskolin-treated slices revealed only subtle changes in synaptic vesicle organization, supporting the supposition that cAMP-mediated increases in mossy fiber synaptic transmission operate via molecular regulation of post-docking processes. However, I observed a two-fold forskolin-induced increase in DCV docking at mossy fiber active zones that implicates neuropeptide signaling as a potential contributing factor involved in presynaptic mechanisms of mossy fiber plasticity.

4.2. Methodological considerations

The main objective of this study was to examine whether the number of morphologically docked synaptic vesicles corresponds to differences in release probability exhibited between Schaffer collateral (weakly facilitating or depressing; phasic) and mossy fiber (strongly facilitating, tonic) synapses in the hippocampus. I asked the question: Do mossy fiber and Schaffer collateral synapses differ not only in the total number of active zones and synaptic vesicles, but also in the organization of vesicles at individual active zones? My results reveal previously unreported ultrastructural characteristics of hippocampal mossy fibers and demonstrate a clear correlation between the availability of docked and primed vesicles and the initial release probabilities of hippocampal mossy fiber and Schaffer collateral synapses. Although this fundamental ultrastructure-function relationship has been examined in a variety of model systems and synapse types, discrepancies between studies have prevented a general consensus (Atwood and Karunanithi, 2002; Eltes et al., 2017; Govind et al., 1994; Holderith et al., 2012; Millar et al., 2002; Schikorski and Stevens, 1997; Xu-Friedman et al., 2001). Nevertheless, I was motivated to revisit this question, particularly since mossy fiber synapses have not previously been scrutinized at the level of resolution permitting accurate discrimination of functionally distinct membrane-proximal vesicle pools. Several important methodological considerations need to be taken into account in order to interpret and compare my data with the results of previous studies obtained using alternative experimental approaches to address synaptic ultrastructure-function relationships. In the following section, I classify and discuss these limitations in the context of sample preparation for electron microscopic analysis, the mode of image acquisition, and of analysis of synaptic ultrastructure.

4.2.1. Sample preparation

In contrast to many previous studies based on the analysis of aldehyde-fixed tissue, my book work was almost entirely founded upon the ultrastructural analysis of tissue that had been rapidly cryo-fixed by HPF in a living, unfixed state. To assess the potential benefits of this approach in the context of synaptic vesicle docking analyses, I compared the effects of aldehyde fixation on synaptic vesicle docking in hippocampal mossy fiber synapses prepared for ultrastructural analysis by i) transcardial perfusion of aldehyde fixative cocktails, ii) immersion of organotypic slice cultures in aldehyde fixative cocktails, and by iii) HPF of organotypic slice cultures.

I found that in comparison to high-pressure frozen organotypic slices, the abundance of membrane proximal and docked vesicles was severely reduced in perfusion fixed tissue. Moreover, I observed that the extent of perturbation correlated with the fixative osmolarity, perhaps reflecting the sensitivity of the synaptic RRP, and presumably of the vesicle fusion apparatus, to osmotic pressure (Bekkers and Stevens, 1995; Rosenmund and Stevens, 1996; Stevens and Tsujimoto, 1995). The osmolarity of aldehyde fixative cocktails typically exceeds that of physiological buffers (Hayat, 1981). In my study, two different fixative cocktails were used: PF1 [4% PFA, 2.5% GA in 0.1 M PB, pH 7.4, 4° C; approximately 1900 mOsm (Hayat, 1981)] and PF2 [2% PFA, 2.5% GA, 2 mM $CaCl_2$ in 0.1 M cacodylate buffer, 37° C; approximately 1200 mOsm (Hayat, 1981)]. Another consideration concerns the speed with which the synaptic ultrastructure is immobilized. Aldehyde fixation via transcardial perfusion occurs relatively slowly because the fixative must diffuse through tissue while cross-linking proteins in an outside-in direction (Hopwood, 1969).

My observation that membrane proximal vesicles are depleted in perfusion-fixed mossy fiber synapses is in line with a previous study comparing the ultrastructural organization of synapses in the somatosensory cortex prepared by transcardial aldehyde perfusion fixation and by HPF of acutely dissected tissue (Korogod et al., 2015). Korogod and colleagues used serial section transmission EM, electron tomography, and focused ion-beam scanning electron microscopy (FIB-SEM) to compare the effects of perfusion fixation of aldehyde cocktails to high-pressure frozen acute slice preparations (Korogod et al., 2015). Using electron tomography, they found a reduction in membrane-proximal synaptic vesicles in reconstructed tomograms of synaptic profiles from perfusion-fixed compared to cryo-fixed tissue (Korogod et al., 2015). I interpret the loss of membrane proximal vesicles in my study as a consequence of aldehyde-triggered vesicle fusion rather than a rapid redistribution from the active zone membrane. This hypothesis is supported by my frequent observation of omega-shaped membrane profiles at active zones in the perfusion-fixed mossy fiber synapses, which is indicative of full-collapse synaptic vesicle fusion. The notion that aldehyde exposure can trigger vesicle fusion was initially proposed by Smith and Reese (Smith and Reese, 1980), who observed that perfusion of different aldehydes: GA, formaldehyde, or crotonaldehyde, caused an increase in post-synaptic endplate potentials recorded from the frog neuromuscular junction, as well as an increase in pits in proximity to the active zone attributed to fusing vesicles (Smith and Reese, 1980). A contrasting view was presented by Rosenmund and Stevens, who recorded EPSCs in dissociated hippocampal neurons and found minimal synaptic vesicle fusion upon fast perfusion of 2% GA (Rosenmund and Stevens, 1997). They found that fixation was rapid and caused a 10% depletion of the functional RRP (Rosenmund and Stevens, 1997), indicating that the speed at which synaptic ultrastructure is immobilized is likely an important factor involved in the effect of aldehydes on docked vesicle pools.

Consistent with this view, the number of docked vesicles at synapses in immersion-fixed and high-pressure frozen slices was comparable at the surface of the tissue (5-11 µm from the tissue surface; Table 30, Figure 21 A-E), whereas the depletion of docked vesicle pools became apparent in synapses located deeper in the tissue (20-22 µm from tissue surface; Table 31; Figure 21 F-I). Since identical aldehyde-based fixative cocktails were used for perfusion and immersion fixation experiments, these data imply that synaptic ultrastructure in surface-exposed synapses is immobilized before the osmotic pressure exerted by the fixatives can manifest and trigger vesicle fusion. However, additional analysis is required to assess whether docked vesicle pools are maintained only at the expense of membrane-proximal, putatively tethered vesicles in surface-exposed and immersion fixed synapses. Regardless, my observations, which appear consistent with the aforementioned study by Rosenmund and Stevens (Rosenmund and Stevens, 1997), identify synapse location and accessibility as important variables affecting the impact of immersion fixation on synaptic ultrastructure.

Taken together, the results of my study emphasize that caution must be exercised when interpreting synaptic ultrastructure in aldehyde-fixed samples and that rapid cryo-fixation is more reliable for the analysis of synaptic vesicle organization at presynaptic active zones. I cannot, however, exclude the possibility that more refined fixative compositions or perfusion protocols may limit the artifactual effects of aldehydes on vesicle organization at individual active zones. Aldehyde fixation will remain an extremely useful means of preserving synaptic ultrastructure for light and electron microscopic analysis, since many brain structures are incompatible with rapid dissection for cryo-fixation (i.e. HPF), and the dissection process itself risks mechanical trauma and anoxia (Korogod et al., 2015; Sosinsky et al., 2008). Protocols involving sequential aldehyde perfusion, dissection, and HPF have therefore been proposed (Sosinsky et al., 2008), the effects of aldehydes on membrane-proximal vesicle pools notwithstanding.

Future work based on my study would be to investigate the spatial distribution of synaptic vesicles within 100 nm of the active zone in mossy fiber synapses from immersion-fixed hippocampal slices to test whether the membrane-proximal, or tethered synaptic vesicle pools change in mossy fiber synapses after immersion fixation. It is possible that in the superficial mossy fiber synapses, the RRP was depleted by 10%, as was observed in the Rosenmund and Stevens study (Rosenmund and Stevens, 1997), however, the loosely docked (LS) or tethered synaptic vesicles observed in mossy fiber synapses at rest may rapidly dock and prime during immersion fixation. To better study the effects of aldehydes on membrane-proximal pools of synaptic vesicles, a full 3D study of synaptic vesicles within 100 nm is necessary. Furthermore, a more thorough study of the diffusion distance of aldehydes is necessary to test whether the distance of fixatives is proportional to the deficits of membrane-proximal synaptic vesicles observed in this study. This could be accomplished by comparing the change in the spatial distribution of synaptic vesicles at mossy fiber synapses closer to blood vessels with those farther away in perfusion-fixed mice.

4.2.2. Limitations of 2D electron microscopy and advantages of 3D electron tomography

Although conceptually simple, the accurate assessment of synaptic vesicle docking is non-trivial. Multiple factors contribute to this: (i) Synaptic vesicles are small and the defining structural feature of a docked synaptic vesicle, i.e. the point of contact between the outer lipid bilayer of a synaptic vesicle and the inner lipid bilayer of the presynaptic membrane, is considerably smaller. (ii) Synaptic membranes are inherently curved. (iii) Electron microscopic images, even from ultrathin sections (20-100 nm thick), represent 2D projections through a volume. (iv) Due to mechanical limitations, plastic sections are difficult to cut thinner than the approximate diameter of a synaptic vesicle (~40 nm). In addition to the sample preparation considerations outlined in the previous section, the accuracy with which vesicle docking is assessed in electron micrographs is dependent on z-resolution. Since z-resolution is defined by plastic section thickness in 2D imaging approaches, including serial section-based 3D EM techniques, certain ambiguities are introduced and it is not possible to determine whether the midline of a synaptic vesicle is contained within the imaged volume. This limitation also excludes the possibility of accurate vesicle volume measurements. Nevertheless, 3D serial section EM analyses offer a larger field-of-view than electron tomography, are compatible for the reconstruction of large tissue volumes, and provide access to important quantitative information, including volume measurements of pre- and postsynaptic compartments and total numbers of active zone release sites (Chicurel and Harris, 1992; Harris and Sultan, 1995; Rollenhagen et al., 2007; Sätzler et al., 2002; Spacek and Harris, 1998; Xu-Friedman et al., 2001).

Electron tomography circumvents this limitation by achieving a higher z-resolution, which is inversely related to the number of images in the tilt-series, the tilt increment separating them, and the pixel spacing used for image acquisition (Koster et al., 1997). The final voxel dimensions achieved in tomographic slices are therefore dependent on the magnification used to acquire the tilt-series and the extent of binning used for weighted back-projection to convert the tilt series into a volume (Koster et al., 1997). Since functionally critical, but morphologically subtle, changes in vesicle organization can manifest within the range of several nanometers (Imig et al., 2014; Siksou et al., 2009a), it is important to resolve the spatial organization of synaptic vesicles as accurately as possible. For example, electron tomographic reconstructions generated in my comparative analyses revealed that mossy fiber synapses not only harbor a lower spatial density of docked vesicles at individual active zones compared to Schaffer collateral synapses, but that they are distinguished by the presence of a prominent membrane-proximal, possibly tethered, pool of synaptic vesicles. The novelty of these observations implies that the combination of good ultrastructural preservation and high resolution imaging is required to dissect functionally relevant structural features of vesicle organization at active zone release sites. Consistent with this view, previously reported differences in the numbers of docked vesicles at facilitating and depressing synapses were generated using electron tomography (Eltes et al., 2017).

In conclusion, I used electron tomography to reconstruct synaptic subvolumes in 3D with an isotropic voxel dimension of about 1.6 nm. In comparison to conventional 2D transmission EM analysis, electron tomography provides a better z-resolution and reveals fine structural changes in the spatial distribution of synaptic vesicles at active zone release sites that have otherwise been elusive.

4.2.3. Vesicle docking criteria

A potential source of discrepancy between studies examining the functional relevance of synaptic vesicle organization is the use of different criteria to classify vesicle docking in an electron micrograph. Relevant questions in such analyses become: (i) How accurately can the 'true' position of a synaptic vesicle be resolved? (ii) How close to the membrane does a vesicle need to be in order to be considered docked? (iii) How do the applied sample preparation techniques (i.e. fixation, dehydration) influence vesicle organization? Indeed, the definition of a morphologically docked synaptic vesicle is often adapted to take the limitations of the applied methodology into account.

For example, in a 3D serial section EM analysis of vesicle docking in perfusion fixed mossy fiber synapses, in order to compensate for the associated limitations in z-resolution in 50-60 nm-thick plastic sections, Rollenhagen and colleagues measured distances between the center of vesicles to the presynaptic membrane and the mean synaptic vesicle radius measured was subtracted to calculate the closest approach (Rollenhagen et al., 2007). Since the total number of synaptic vesicles within 60 nm of the active zone membrane (~1200 vesicles) were found to correlate to the number of vesicles predicted to fuse upon depletion of the RRP as assessed by presynaptic capacitance recordings (~1400 vesicles; Hallermann et al., 2003), these vesicles were classified as belonging to the putative RRP (Rollenhagen et al., 2007). However, this putative RRP would include all vesicles within 40 nm of the active zone membrane resulting in an overestimation of docked synaptic vesicles.

More stringent docking criteria have been used in other 3D serial section EM studies. In an analysis of synaptic vesicle docking in the *stratum oriens* of the CA3 from juvenile rat hippocampi, Holderith and colleagues reconstructed entire active zones from serial 20 nm-thick sections (Holderith et al., 2012). For a synaptic vesicle to be considered docked in this study, the distance between the middle of the lipid bilayers on the presynaptic membrane and synaptic vesicle was required to be less than 5 nm (Holderith et al., 2012). Since my electron tomographic analyses reveal lipid bilayers to be approximately 4 nm thick from center-to-center of inner and outer leaflets, the criteria used by Holderith and colleagues, coupled with the very thin sections imaged, provides a stringent analysis of the number of docked synaptic vesicles (Holderith et al., 2012).

In the present study and in previous work from our group using HPF, AFS, and electron tomography, docked vesicles were defined as those with no measurable distance between the synaptic vesicle membrane and the active zone membrane (Imig et al., 2014). In both studies, tomograms were acquired at 30,000x magnification and a 3x binning factor was used when reconstructing synaptic subvolumes, generating tomographic slices with isotropic voxel dimensions of x, y, and z = ~1.6 nm (Imig et al., 2014). Morphologically docked vesicles were reported as vesicles within 0-2 nm of the active zone membrane. This study of Schaffer collateral synapses indicated that docked vesicles classified according to this criterion represent morphological correlates of functionally primed and fusion-competent vesicles (Imig et al., 2014). Docked vesicle pools within 0-2 nm of the membrane are massively reduced in priming and SNARE protein deficient genetic mutants (Imig et al., 2014) and the extent of reduction is highly comparable to published reductions in sucrose-evoked measurements of RRP size in respective mutants (Arancillo et al., 2013; Augustin et al., 1999; Bronk et al., 2007; Schoch et al., 2001; Varoqueaux et al., 2002; Washbourne et al., 2002). Although our workflow combining HPF and AFS circumvents many potential limitations of aldehyde fixation and room-temperature dehydration, the tissue is nevertheless dehydrated, albeit at sub-zero temperatures, and heavy metals are deposited on the lipid bilayers to enhance membrane contrast. Consequently, the possibility that these procedures could induce subtle changes in the distribution of vesicles at active zones, or even occlude the detection of very small gaps between vesicles and the presynaptic membrane, should be taken into consideration.

An alternative perspective of how synaptic vesicles interact with the presynaptic membrane has evolved with the technological development of cryo-EM, which allows the ultrastructure of synapses to be viewed in a frozen-hydrated state (Fernández-Busnadiego et al., 2010, 2013; Lučić et al., 2005; Schrod et al., 2018; Tao et al., 2018b; Zuber and Lučić, 2019). Studies employing cryo-electron tomography to resolve the active zone organization of plunge-frozen synaptosome preparations demonstrated that direct synaptic vesicle contact with the plasma membrane was only rarely observed (Fernández-Busnadiego et al., 2010). Rather, active zone proximal vesicles were connected to the membrane by multiple short filaments (<5 nm)

(Fernández-Busnadiego et al., 2010). The authors interpreted vesicles in this state as potentially primed and analogous to the docked vesicles in direct contact observed in heavy metal contrasted and dehydrated preparations (Fernández-Busnadiego et al., 2010). Although aspects of the synaptosome preparation protocol likely induce some degree of structural reorganization in presynaptic vesicle pools, more recent studies corroborate several observations made in synaptosomes using cultured neurons (Schrod et al., 2018; Tao et al., 2018a, 2018b). In these experiments, dissociated neuron cultures were maintained on EM grids and vitrified by plunge freezing prior to cryo-electron tomographic analysis (Schrod et al., 2018; Tao et al., 2018b). These studies of intact synaptic boutons confirmed that direct vesicle-membrane contact was rare, and that vesicles closest to the active zone membrane were situated at a small distance spanned by multiple, short filamentous tethers (Schrod et al., 2018; Tao et al., 2018b). The molecular identity of the short filaments remains unknown and future experiments quantifying their abundance in appropriate genetic null mutants (i.e. synapses lacking SNARE components or priming proteins), will be informative.

In summary, it is important to understand the advantages and limitations of the employed methodologies when assessing synaptic vesicle docking. In my study, I used 3D electron tomography which resulted in isotropic voxel dimensions compatible to the accurate assessment of synaptic vesicle docking. I set a stringent docking criterion—no measurable distance between the membranes of synaptic vesicles and the presynaptic active zone. While electron tomography of plastic-embedded samples provides better 3D resolution than 2D transmission EM, cryo-electron tomography images biological samples with no additives such as those used for plastic embedding. An additional comparison between the spatial organization of synaptic vesicles in synapses imaged with cryo-electron tomography and electron tomography of plastic-embedded synapses would be highly informative.

4.3. The RRP and morphologically docked vesicles

The question of whether morphologically docked synaptic vesicles at a given synapse can predict differences in synaptic release probability has been contentious (Atwood and Jahromi, 1978; Eltes et al., 2017; Holderith et al., 2012; Neher and Brose, 2018; Xu-Friedman and Regehr, 2003; Xu-Friedman et al., 2001). At the core of this exploration comes another question: do docked synaptic vesicles comprise the RRP of primed, fusion-competent synaptic vesicles? The fact that loss of Munc13-1 and -2 abolishes both docking and priming indicates that the number of docked vesicles (Figure 11, Figure 15, and Figure 19) serves as a reliable proxy for the number of molecularly primed vesicles (Imig et al., 2014)

It is also clear that in response to a given action potential, only a fraction of docked and therefore primed synaptic vesicles will fuse. The RRP is a vague concept that usually requires some degree of activity to extrapolate the approximate number of docked and primed synaptic vesicles. With HPF, AFS, and electron tomography of mossy fiber synapses, I defined the RRP of morphologically docked synaptic vesicles from these synapses at rest. I compared my findings to past functional studies that used presynaptic capacitance recordings of mossy fiber synapses to estimate the RRP of mossy fiber boutons (Hallermann et al., 2003; Midorikawa and Sakaba, 2017). Our theoretical RRP estimates fell within the reported range of 400-1400 synaptic vesicles (Hallermann et al., 2003; Midorikawa and Sakaba, 2017). I calculated, based on the average docked vesicle density, that approximately 320 synaptic vesicles, 37 giant vesicles and 15 DCVs are docked and primed in a given mossy fiber bouton at rest (Table 12). The authors of the previous work describing RRP estimates from presynaptic capacitance recordings of mossy fiber boutons acknowledge that they do not account for DCV fusion nor exocytosis from filopodia that extend from mossy fiber boutons and form synapses on inhibitory interneurons (Acsády et al., 1998; Hallermann et al., 2003; Midorikawa and Sakaba, 2017). The neglect of giant vesicles' potential contributions to induced capacitance changes at hippocampal mossy fiber synapses may lead to an overestimation of the RRP (Midorikawa and Sakaba, 2017; Rollenhagen et al., 2007).

I found that the total docked vesicle pool in mossy fiber synapses is several hundred vesicles. However, the number of vesicles that seem to fuse in response to an AP appears to be much smaller (Jonas et al., 1993; Lawrence et al., 2004). This implies a heterogeneous release probability among the docked and primed synaptic vesicle pool. Both fast and slow components of synaptic transmission have been described and the contribution to each component is provided by vesicles of certain release probabilities. This was observed in calyx of Held synapses and is partially due to the synaptic vesicle-VGCC coupling distance (Chen et al., 2015). Vesicles within the fast pool are coupled to VGCCs at a distance of ~16 nm whereas vesicles in the slow pool were estimated to reside 30 to 100 nm from the calcium channels (Chen et al., 2015; Neher, 2015). A recent study looked at the freeze-fracture replica immunolabelling of VGCC distribution in two synapses in the cortex with high and low release probability (Rebola et al., 2019). They found the opposite correlation of VGCC density to release probability in that synapses with lower release probability had a higher density of VGCCs and synapses with a higher release probability had a lower abundance of VGCCs (Rebola et al., 2019). Rebola and colleagues also found that synapses with a lower release probability had a larger calcium channel-synaptic vesicle coupling distance than synapses with a higher release probability (Rebola et al., 2019). Mossy fiber synapses have a loose calcium channel-synaptic vesicle coupling distance of approximately 70 nm, and although the total number of calcium channels has been estimated to be about 23 per active zone (Vyleta and Jonas, 2014), the actual distribution of calcium channels remains to be determined.

Another important consideration associated with presynaptic capacitance concerns control of presynaptic intracellular calcium concentrations and the distance calcium ions travel within the bouton during a given step-depolarization protocol. Indeed, the addition of calcium buffers to intracellular recording solutions, which compete with endogenous buffers within the presynaptic terminal, can profoundly influence RRP estimates. For example, in work from Hallermann and colleagues, the presynaptic calcium buffer concentration was half of that used in the work published by Midorikawa and Sakaba (0.26 mM EGTA, Hallermann et al., 2003; 0.5 mM EGTA, Midorikawa and Sakaba, 2017). The RRP estimate extrapolated from the study by Hallermann and colleagues likely contained docked and primed vesicles as well as membrane-proximal vesicles (loosely-docked) that rapidly primed upon sustained calcium influx, therefore inflating the RRP estimate (Hallermann et al., 2003). For this reason, I incorporated loosely-docked synaptic vesicles, synaptic vesicles within 0-40 nm of the active zone, into my RRP estimate (Table 12). The theoretical change in membrane capacitance when including loosely-docked synaptic vesicles is within the range of the upper estimates from presynaptic capacitance measurements (Table 12; Table 13; Hallermann et al., 2003).

My study supports the finding that morphologically docked synaptic serve as a reliable proxy for the RRP of a given synapse (Imig et al., 2014; Siksou et al., 2009a). However, in the tomograms of Schaffer collateral and mossy fiber synapses reconstructed and analyzed in this study, each tomogram contained a fraction of an entire active zone. Therefore, to make RRP estimates per active zone and per bouton, I relied on active zone areas reported from past morphological studies (Chicurel and Harris, 1992; Rollenhagen et al., 2007). Ideally, electron tomograms of serial semi-thin (200 nm-thick) sections could be used to reconstruct entire active zones at both Schaffer collateral and mossy fiber synapses to provide a more accurate estimate of the RRP.

4.3.1. Limitations of RRP estimates

The RRP has been historically assessed by functional means and corresponds to the number of vesicles that fuse with the synapse in response to strong, vesicle-depleting stimuli (Kaeser and Regehr, 2017; Neher, 2015). Several approaches assess the RRP of a particular neuron or synapse: high-frequency stimulation (Schneggenburger et al., 1999), presynaptic membrane capacitance recordings (Neher and Marty, 1982), optical approaches including FM1-43 dye uptake (Rizzoli and Betz, 2004; Schikorski and Stevens, 2001) and phlourins (Ariel and Ryan, 2010), and perfusion of hypertonic sucrose solutions (Rosenmund and Stevens, 1996).

Most commonly, the RRP for synapses in acute slice preparation is measured by recording action potential-evoked EPSCs during high-frequency stimulation paradigms to compare the responses to the quantal content of a single synaptic vesicle from spontaneous fusion events and back extrapolate the number of vesicles that form the RRP (Schneggenburger et al., 1999; Thanawala and Regehr, 2013). However, it is an indirect measurement of vesicle fusion as it depends on the detection of neurotransmitter molecules by postsynaptic receptors. Moreover, during long high-frequency stimulation trains the RRP is constantly refilled (Neher, 2015).

Optical approaches offer a sensitive way to study the RRP in neurons and to determine which vesicles participate in the vesicle cycle. For example, a key study from Rizzoli and Betz used photoconversion of FM1-43 dye uptake via endocytosis after pool-depleting simulation of the frog neuromuscular junction to assess the spatial distribution of photoconverted synaptic vesicles with EM (Rizzoli and Betz, 2004). Another method is based on the use of phlourins, pH-sensitive fluorophores coupled to synaptic vesicle proteins, which become excitable and emit light upon vesicle fusion and quenched upon re-acidification of the synaptic vesicle (Ariel and Ryan, 2010). This technique enables monitoring synaptic vesicle fusion live and therefore offers good temporal resolution.

Presynaptic capacitance measurements are a direct method to monitor the fusion of vesicles with the plasma membrane, however, they are only applicable for the study of large synaptic boutons (Neher and Marty, 1982). Using this method, the change in membrane capacitance in response to strong stimulation paradigms is measured and directly related to the addition of membrane to the presynaptic terminal via synaptic vesicle fusion or removal of membrane through endocytosis. Although this approach has extremely high temporal resolution, it cannot distinguish between exo- or endocytosis-mediated membrane capacitance changes during stimulation or the morphological nature of the membrane trafficking events. Moreover, in large terminals harboring multiple active zone release sites (i.e. hippocampal mossy fiber boutons) whole-bouton capacitance changes are insensitive at the level of individual active zone release sites.

A common method used to measure the RRP in mixed neuron cultures is by perfusing hypertonic sucrose solution over the cell and recording the postsynaptic response (Rosenmund and Stevens, 1996). This technique is especially powerful in combination with low-density neuron culture systems, in which neurons form synapses onto themselves (autapses), because these autaptic cultures allow a very standardized experimental system to study the molecular mechanisms underlying synaptic function in individual neurons (Bekkers and Stevens, 1991). Hypertonic sucrose solutions cause the fusion of synaptic vesicles with the plasma membrane, which is believed to be independent of calcium influx and potentially caused by shrinkage of the presynaptic bouton. However, the exact mechanism of hypertonic sucrose-induced synaptic vesicle fusion is poorly understood.

The aforementioned techniques to determine RRP content in neurons require a degree of evoked fusion of synaptic vesicles. Factors such as calcium-dependent priming, short-term plasticity, and postsynaptic receptor saturation and sensitization can influence the RRP estimate. With HPF, freeze substitution, and electron tomography, I was able to characterize morphologically docked synaptic vesicles from synapses at rest.

4.4. Morphological correlates of mossy fiber facilitation

Release probability and short-term plasticity are dependent upon multiple factors, including calcium channel-sensor coupling distance (Vyleta and Jonas, 2014), shape of the action potential (Geiger and Jonas, 2000), calcium buffers (Blatow et al., 2003; Dumas et al., 2004; Vyleta and Jonas, 2014), type of calcium sensor (Jackman et al., 2016), and accessibility to energy sources (i.e. mitochondria) (Kwon et al., 2016).

Despite comparable numbers of membrane-proximal synaptic vesicles within 40 nm of the active zone membrane, I found fewer morphologically docked and primed synaptic vesicles at individual mossy fiber active zones compared to Schaffer collateral synapses from the same slice. These data indicate that differences in the availability of docked and primed vesicles could, in addition to other factors, co-determine initial release probability. Past studies examined whether release probability is dependent on the availability of morphologically docked synaptic vesicles, however the conclusions did not lead to a general consensus (Eltes et al., 2017; Holderith et al., 2012; Millar et al., 2002; Schikorski and Stevens, 1997; Xu-Friedman et al., 2001). This question was worth revisiting in light of some limitations from past studies that can be circumvented with different techniques (i.e. cryo-fixation and electron tomography).

My findings are consistent with those reported by Eltes and colleagues that synapses with a low release probability (facilitating, tonic) harbor fewer docked synaptic vesicles than synapses with higher release probability (depressing, phasic) (Eltes et al., 2017). Along the same lines, Schikorski and Stevens found that the variability in the number of docked synaptic vesicles in the rat CA1 accounted for the heterogeneity in synaptic release probability and concluded that the number of docked vesicles contributes to synaptic release probability (Schikorski and Stevens, 1997). Conversely, my results differ from studies in the rodent cerebellum where synapses from parallel fibers (low release probability, facilitating) and synapses from climbing fibers (high release probability, depressing) harbored similar numbers of docked synaptic vesicles (Xu-Friedman et al., 2001). Further, in hippocampal associational/commissural synapses the number of docked synaptic vesicles and the release probability positively correlated with active zone area (Holderith et al., 2012).

In contrast to Schaffer collateral synapses, mossy fiber synapses at rest had a second pool of membrane-proximal vesicles 5-20 nm from the active zone. I interpret the membrane-proximal accumulation of vesicles in mossy fibers as structural evidence of a tethering step preceding synaptic vesicle docking/priming. Although undetected in wild-type Schaffer collaterals in the present study, the membrane-proximal accumulation of vesicles is highly reminiscent of the synaptic vesicle organization observed in Schaffer collateral synapses lacking key priming proteins, such as Munc13s, and core components of the neuronal SNARE complex, such as SNAP-25 and synaptobrevin-2 (Imig et al., 2014). This indicates that membrane-proximal vesicle tethering is not unique to mossy fibers *per se*, but implies that molecular processes responsible for forming the tethered pool operate differently in Schaffer collateral and mossy fiber synapses. Differences in molecular processes could involve a limited copy number of any of the aforementioned priming and SNARE proteins that, upon genetic deletion, manifest a membrane-proximal accumulation of synaptic vesicles. Specialized vesicle tethering mechanisms have evolved to support the transmitter release properties, and behaviors of distinct synapse types have been shown in specialized synapses such as invertebrate neuromuscular junctions and mammalian ribbon synapses (Hallermann and Silver, 2013).

My findings support a previously proposed vesicle tethering mechanism (Hallermann and Silver, 2013) for the first time in mossy fiber synapses. Synaptic vesicles are thought to loosely tether to the plasma membrane close enough for interaction to occur between v-SNAREs and t-SNARE (Neher and Brose, 2018). A tightening of the SNARE complex assembly results in morphological docking of synaptic vesicles to the plasma membrane mediated by Munc13 and CAPS priming molecules (Neher and Brose, 2018). These steps represent a loose and then tight synaptic vesicle docking state (Neher and Brose, 2018). The generation of a LS synaptic vesicle pool has been shown in dissociated mouse hippocampal neurons to rely on the calcium sensor synaptotagmin-1 (Chang et al., 2018). My finding of a membrane proximal pool of synaptic vesicles in mossy fiber synapses indicates the morphological correlate of tethered and therefore LS synaptic vesicles. The relationship between these tethered vesicles and the hypothesized LS state in tonic mossy fiber synapses requires further investigation, both in terms of its reliance of synaptotagmin-1, and in terms of how quickly the tethered pool is formed or depleted. These questions could be addressed by performing flash-and-freeze experiments-coupled with 3D ultrastructural analysis in synaptotagmin-1 mutant hippocampal slice cultures analogous to the study in dissociated neuron culture by Chang and colleagues (Chang et al., 2018).

Although I did not directly quantify filamentous material in my tomograms, the distance at which the membrane-proximal pool accumulated in wild-type mossy fiber synapses in the present study, and in priming-deficient Schaffer collateral synapses (Imig et al., 2014) is highly comparable to the length of long, single tethers described in cryo-electron tomographic reconstructions of frozen-hydrated synaptosomes (Fernández-Busnadiego et al., 2010, 2013). Since the molecular identity of vesicle tethers remains to be clarified, it is difficult to speculate which proteins are specifically responsible for generating a prominent membrane-proximal pool in wild-type mossy fibers at rest. Whereas synaptosomes isolated from RIM1α-deficient mice exhibited a perturbation of filamentous active zone material, it is unlikely that tethering is mediated by RIM1α alone. Another candidate is bassoon, an active zone protein that has been implicated in rapid RRP replenishment in the cerebellum during high synaptic activity (Hallermann et al., 2010) and is necessary for the proper maturation of active zones in hippocampal mossy fiber synapses (Lanore et al., 2010).

I hypothesize that the tethered vesicle pool I observed at mossy fiber active zones is ideally situated to resupply the docked and primed pool of vesicles during sustained activity. Moreover, I propose that a low ratio of docked to tethered vesicles, as I observed in hippocampal mossy fibers, may serve as a structural feature distinguishing facilitating ("tonic") synapses. This view is compatible with recently proposed theories postulating that "tonic" and "phasic" synapses have different ratios of tightly (TS) and loosely (LS) docked vesicle states (Neher and Brose, 2018). In the future, additional experiments examining the time course of docked and tethered pool depletion during induced short- and long-term plasticity regimes are necessary to test these hypotheses.

A direct visualization and quantification of mossy fiber tethers by cryo-electron tomography would be informative, particularly in combination with genetic perturbations. Nevertheless, to perform such experiments in a tissue context would be exceptionally technically challenging, requiring a combination of HPF and the subsequent generation of lift-out FIB lamella and ultimately tomographically reconstructed under cryo conditions (Mahamid et al., 2015; Schaffer et al., 2019). Additional technical challenges would need to be addressed, including how to identify mossy fiber boutons in the vitrified sample, and how to identify active zones in the ideal orientation for tomography.

4.5. Synaptic vesicles, giant vesicles, and dense core vesicles at mossy fiber active zones

Consistent with previous publications, my data revealed considerable morphological heterogeneity in the presynaptic vesicle pools at mossy fiber active zones, which comprised clear core synaptic vesicles (diameter range 38-59 nm), giant clear core vesicles, (diameter range 60-120 nm), and DCVs (diameter range 45-100 nm). However, my data represent the first systematic morphometric analysis of their dimensions and relative spatial distributions with respect to active zone release sites. Moreover, I demonstrate for the first time clear ultrastructural evidence that all three vesicle types dock in direct physical contact with the active zone membrane in a Munc13-dependent manner.

4.5.1. Giant vesicles

Although the presence of giant vesicles in mossy fiber boutons has been described (Henze et al., 2002; Laatsch and Cowan, 1966; Rollenhagen et al., 2007), the origin, cargo, and functional implications of this fascinating organelle remain enigmatic. Several lines of evidence support the hypothesis that giant vesicles containing glutamate neurotransmitter cargo contribute to glutamatergic signaling: i) electron micrographs revealed giant vesicles in proximity to mossy fiber active zones (Henze et al., 2002b; Laatsch and Cowan, 1966; Rollenhagen et al., 2007); ii) large amplitude "giant" mEPSCs recorded from postsynaptic CA3 pyramidal neurons were demonstrated to be monoquantal (Henze et al., 2002b); iii) gamma-radiation lesions ablating the DG abolished giant mEPSCs recorded from CA3 pyramidal neurons (Henze et al., 1997). My data expands upon these studies by demonstrating that the proportion of docked giant vesicles is highly comparable to the proportion of giant mEPSCs recorded from pyramidal cells. Although direct evidence that giant vesicles contain glutamate, or possess vesicular glutamate transporters, is still lacking. However, my observation that giant vesicle docking is abolished by deletion of Munc13 priming proteins indicates that they likely harbor at least some of the vesicular molecules required for evoked fusion.

In the scenario that if giant vesicles, though capable of docking, were actually incapable of fusing at the active zone membrane, it is conceivable that they limit access of "normal" synaptic vesicles to release sites. This hypothesis is not supported by my analyses, which failed to detect a correlation between docked vesicle numbers and the proportion of giant vesicles (data not shown).

Multiple questions remain to be answered. Do giant vesicles fuse? Do they have the same vesicular release probability as a "normal" synaptic vesicle? Where and how are giant vesicles generated? Are these organelles unique to mossy fiber synapses? Several forms of activity-dependent membrane retrieval operate in synapses that could generate large vesicular structures, including compound fusion (He et al., 2009), ultrafast endocytosis (Delvendahl et al., 2016; Watanabe et al., 2013b, 2013a), and bulk membrane retrieval (Cousin, 2009). Indeed, in the calyx of Held, large neurotransmitter-filled giant vesicles have been shown to result from activity-induced compound fusion of synaptic vesicles (He et al., 2009). Ultrafast endocytosis mechanisms generate large spherical endocytic intermediates in response to action potential-evoked release from small glutamatergic synapses from cultured hippocampal neurons (Watanabe et al., 2013a). More recently, functional analyses of endocytosis kinetics indicate that ultrafast modes of endocytosis also operate in other central synapses (Brockmann and Rosenmund, 2016; Delvendahl et al., 2016). Bulk membrane retrieval has only been observed in synapses during high activity states (Cousin, 2009) and is unlikely to be the origin of giant vesicles in mossy fiber synapses. It is also possible that giant vesicles result from de-granulation DCVs in a process comparable to that described for large DCVs in chromaffin cells (Shin et al., 2018). Although I occasionally observed filamentous electron dense material in the lumen of giant vesicles in the mossy fiber synapse, giant vesicles in mossy fibers persisted following pharmacological or genetic silencing in Munc13-deletion mutants. Since both synaptic and DCV fusion at synaptic active zones is severely decreased upon loss of Munc13 priming proteins (Augustin et al., 1999; van de Bospoort et al., 2012; Varoqueaux et al., 2002), my data indicate that at least a considerable subpopulation of giant vesicles are generated via activity-independent mechanisms. Nevertheless, further analyses of ultrastructural changes induced during defined activity states are required to establish the contribution of compensatory endocytosis to large vesicle pools in mossy fiber synapses.

Alternative options include a form of constitutive membrane retrieval, which operates preferentially in mossy fiber synapses, or anterograde axonal trafficking of giant vesicles from the soma. I found evidence of giant vesicles in granule cell axonal projections in the CA3 from acute slice preparations alongside the other vesicle types, synaptic vesicles, and DCVs. This could indicate anterograde trafficking from the soma, however the transport direction of these vesicles in plastic embedded samples cannot be determined. Moreover, the purpose of these precursor giant vesicles and whether they are molecularly equipped to contribute to synaptic transmission is unclear. If a proportion of giant vesicles are morphological correlates of membrane retrieval, adaptor protein-3, an endocytosis-associated protein, would be one molecular candidate to further investigate this line of inquiry as it has a specific influence on mossy fiber endocytosis and synaptic vesicle dynamics (Evstratova et al., 2014; Scheuber et al., 2006).

The release of large quanta from giant vesicles has potential implications for mossy fiber function and several important questions remain open. For example, do giant vesicles fuse with the same release probability as other synaptic vesicles? Does quantal release from giant vesicles contribute to mossy fiber facilitation? These could be addressed by measuring mEPSCs in CA3 pyramidal neurons with increasing external calcium concentrations to determine whether the spontaneous fusion of giant vesicles changes in a calcium-dependent manner. In a previous study, the frequency of giant monoquantal events on CA3 pyramidal neurons did not increase in a calcium-dependent manner (Henze et al., 2002b). It has been further shown that vesicles with high membrane curvature, such as small synaptic vesicles, promote more lipid mixing than low-curvature vesicles, such as giant vesicles (Malinin et al., 2002). These findings indicate already that giant vesicles intrinsically have a lower fusogenicity than small synaptic vesicles.

Future directions to explore in terms of giant vesicles would include determining what proportion of giant vesicles represent endocytic intermediates generated by constitutive pathways. This question can be approached in several ways: (i) uptake assays using cell-impermeable fluorescent dyes, photoconvertable dyes, or electron dense particles; (ii) inhibition of endocytosis using pharmacological application of dynamin inhibitors; and (iii) changes in giant vesicle abundance/morphological characteristics in response to defined synaptic activity regimes.

In the case of uptake assays, ferritin (Farrant, 1954; Watanabe et al., 2013a), phluorins (Ariel and Ryan, 2010), and FM1-43 dyes (Branco et al., 2010; Rizzoli and Betz, 2004) have been used to investigate endocytosis or endocytic by-products in synapses by being captured in the membrane invagination during endocytosis. These methods are usually ideal for cell monolayers as is the case in autapses and dissociated hippocampal neuron culture (Ariel and Ryan, 2010; Branco et al., 2010; Watanabe et al., 2013a).

In the case of inhibiting endocytosis, DYNAsore and DYNgo are inhibitors of dynamin, a molecule necessary for pinching endocytosed vesicles to detach them from the presynaptic membrane (Daniel et al., 2012; Mccluskey et al., 2013). The relative abundance of giant vesicle-sized invaginations could be compared to the relative abundance of giant vesicles within a defined distance from the active zone. Such an experiment could also provide the locations of endocytosis in the synapse, though previous studies in cultured hippocampal neurons have demonstrated that ultrafast endocytosis occurs at peri-synaptic sites, at the periphery of the active zone (Watanabe et al., 2013a). One caveat to dynamin inhibition in slice cultures is that dynamin has a role in other membrane trafficking events within the entire cell, and cytotoxicity of the dynamin inhibitors could partially occlude the findings (McMahon and Boucrot, 2011).

4.5.2. Pharmacological enhancement of release probability

Mossy fiber release probability is sensitive to changes in presynaptic cAMP concentration (Fykse et al., 1995; Steuer Costa et al., 2017; Weisskopf et al., 1994). As previously mentioned, I applied forskolin to cultured hippocampal slices to enhance mossy fiber release probability by initiating a signaling cascade involving forskolin-mediated activation of AC1, increased intracellular production of cAMP, and activation of PKA. Forskolin-mediated elevations of cAMP and activation of PKA are used to induce a chemical form of LTP (Fernandes et al., 2015; Huang et al., 1994a; Villacres et al., 1998; Weisskopf et al., 1994). My analyses of forskolin-potentiated mossy fiber boutons reveal only a mild increase in the number of docked synaptic vesicles at the active zone membrane. These findings are in line with previous findings in mossy fiber synapses that forskolin-induced enhancement of synaptic transmission has little effect on the functional RRP measured via electrophysiological approaches (Midorikawa and Sakaba, 2017). These findings indicate that the downstream target of PKA likely acts at a post docking and priming step. Multiple downstream phosphorylation targets of PKA have been identified that also contribute to the regulation of the vesicular release machinery. These include complexins (Cho et al., 2015), rab3A (Castillo et al., 1997; Geppert et al., 1997), and synaptotagmin-12 (Kaeser-Woo et al., 2013).

Complexins are a molecular candidate implicated in mossy fiber LTP (Gibson et al., 2005). Cho and colleagues demonstrated that complexin is phosphorylated by PKA and increases spontaneous synaptic vesicle fusion after tetanus stimulation in the *Drosophila* neuromuscular junction (Cho et al., 2015). Although genetic deletion of complexin-2 caused a reduction in mossy fiber LTP, forskolin-induced potentiation was unaffected (Gibson et al., 2005). Previous ultrastructural analyses in Schaffer collateral synapses from complexin-1, -2, and -3 triple knock-out support the idea that complexin acts on vesicular release machinery downstream of synaptic vesicle docking (Imig et al., 2014).

In mossy fiber synapses, rab3A is not essential for paired-pulse facilitation, a form of short-term plasticity (Castillo et al., 1997). However, genetic deletion of rab3A from mossy fiber synapses ablated mossy fiber LTP (Castillo et al., 1997). Conversely, forskolin-induced mossy fiber potentiation was intact in rab3A knock-out animals (Castillo et al., 1997). Rab3a is not directly phosphorylated by PKA but has strong interaction with two PKA targets at the presynapse, RIM and rabphillin (Castillo et al., 1997; Fykse et al., 1995). Castillo and colleagues propose that rab3a is an effector downstream of PKA and, due to interactions with RIM and rabphillin, has a role in controlling the calcium influx during LTP (Castillo et al., 1997).

Synaptotagmin-12 is phosphorylated by PKA and co-localizes with synaptotagmin-1 on synaptic vesicles in mossy fiber synapses (Kaeser-Woo et al., 2013; Maximov et al., 2007). Although synaptoagmin-12 is not a calcium-sensing synaptotagmin, it is activated via cAMP (Kaeser-Woo et al., 2013; Pang and Südhof, 2010). Synaptotagmin-12 interacts with synaptotagmin-1, inhibiting its function as a calcium sensor, however once phosphorylated by PKA, synaptotagmin-12 dissociates from synaptotagmin-1 thus increasing vesicular release probability (Kaeser-Woo et al., 2013; Maximov et al., 2007).

PKA phosphorylation can also inhibit synaptic proteins, as in the case of tomosyn. Tomosyn prevents SNARE complex assembly by binding syntaxin (Baba et al., 2005). Phosphorylation of tomosyn diminishes its interaction with syntaxin (Baba et al., 2005; Ben-Simon et al., 2015; Fujita et al., 1998; Hatsuzawa et al., 2003), making more syntaxin available for SNARE complex formation at the active zone membrane. Genetic knockdown of tomosyn increases release probability in hippocampal mossy fiber synapses, causing changes to short-term plasticity and occluding LTP (Ben-Simon et al., 2015).

In summary, presynaptic cAMP signaling leads to enhanced synaptic transmission from mossy fiber synapses (Huang et al., 1994a; Villacres et al., 1998; Weisskopf et al., 1994). Pharmacological application of forskolin works on this signaling pathway by enhancing presynaptic cAMP. I tested whether increasing synaptic efficacy at mossy fiber synapses in slice culture increase the number of docked vesicles. However, no differences were observed between forskolin and control or with DCG-VI-treated slices. My findings support the hypothesis that the enhancement of synaptic transmission by cAMP signaling pathways works downstream of synaptic vesicle docking and priming.

4.5.3. DCVs in mossy fiber synapses

The present study found that DCVs dock directly at the active zone in mossy fiber synapses at rest in acute and slice culture preparations. This finding is in contrast to analogous studies in Schaffer collateral synapses from past and present studies (van de Bospoort et al., 2012; Farina et al., 2015; Imig et al., 2014; Siksou et al., 2009a). In the absence of Munc13 priming proteins, DCV docking, as well as all vesicle docking, at mossy fiber active zones was completely abolished and a 2.5-fold increase in DCV abundance was observed within 100 nm of the active zone membrane. Although large DCV docking was unaffected by Munc13-deletion in chromaffin cells (Man et al., 2015), these data are consistent with the finding that Munc13-1 is required for synaptic DCV exocytosis in mammalian neuron cultures (van de Bospoort et al., 2012).

While strong stimulation is necessary to trigger DCV release in cultured neurons (van de Bospoort et al., 2012), I found that mossy fiber synapses at rest harbor docked DCVs and, in the absence of Munc13 priming proteins, DCVs accumulated at the active zone which could indicate that DCV fusion (i.e. neuropeptide release) at mossy fiber synapses occurs to some degree at a basal level. Accordingly, bassoon mutant mice exhibit an accumulation of BDNF- and enkephalin-containing DCVs in the presynaptic terminals of mossy fiber synapses (Dieni et al., 2012, 2015), concurrently, in dissociated mouse hippocampal neurons, bassoon mutants exhibit a reduction in the number of fusion-competent vesicles (Altrock et al., 2003). Both enkephalin and BDNF are involved in synaptic plasticity of the mossy fiber-CA3 synapse (Derrick et al., 1992; Dieni et al., 2012; Li et al., 2010). BDNF contributes to mossy fiber potentiation in a transactivation mechanism that leads to increased synaptic transmission, however the retrograde signaling molecules are still unknown (Huang et al., 2008). Another neuropeptide, enkephalin, binds presynaptic μ-opioid receptors and is involved in frequency-dependent enhancement of mossy fiber-CA3 synaptic transmission during LTP (Derrick et al., 1992). Taken together, these data indicate that synaptic vesicles and neuropeptide-containing DCVs share overlapping molecular mechanisms that operate in steps preceding fusion at mossy fiber active zones.

Alternatively, the accumulation of DCVs may reflect a homeostatic mechanism triggered in response to the loss of network activity in Munc13-deficient slices. In support of this hypothesis, chronic pharmacological silencing of cultured mammalian neurons treated for 48 hours with TTX led to an accumulation of DCVs in both inhibitory and excitatory synapses (Tao et al., 2018a). Furthermore, Tao and colleagues found more membrane-proximal DCVs after chronic TTX treatment (Tao et al., 2018a). Additionally, the researchers found evidence of DCV fusion at the active zone as well as at non-synaptic sites following 48 hours of TTX treatment (Tao et al., 2018a). Tao and colleagues postulated that DCV accumulation in chronically silenced neurons is involved in homeostatic metaplasticity by transporting active zone material to the presynaptic membrane (Tao et al., 2018a). Homeostatic metaplasticity is the compensatory mechanism of a neuron to enhance or diminish synaptic plasticity (i.e. LTP or long-term depression) (Abraham, 2008). In this case, Tao and colleagues propose that additional active zone material is trafficked to chronically silenced synapses to increase the active zone scaffolding material and thus increase the number of release sites at the synapse as has been previously described (Shapira et al., 2003; Sorra et al., 2006; Tao et al., 2018a). However, the comparable size of active zones reconstructed in Munc13-deficent and control mossy fiber synapses in the present study, and the lack of DCV accumulation in Munc13-deficient Schaffer collateral synapses (Imig et al., 2014) appear inconsistent with this notion. To address this, future studies should investigate whether BDNF- or enkephalin-positive DCVs accumulate, whether the DCVs contain active zone molecules such as piccolo or bassoon (Maas et al., 2012; Shapira et al., 2003; Tao-Cheng, 2007), or whether active zones areas increase in mossy fiber synapses in Munc13-deficient slices.

My findings indicate Munc13s facilitate DCV docking and priming at mossy fiber active zones. Past studies show that Munc13-1 is necessary for mossy fiber LTP (Yang and Calakos, 2011). Neuropeptides released from mossy fiber boutons modulate synaptic transmission by modulation of pre- and postsynaptic targets (Chavkin et al., 1983; McQuiston and Colmers,

1996; Salin et al., 1995; Sherwood and Lo, 1999; Weisskopf et al., 1993). I speculate that DCV fusion mediated by Munc13s modulate synaptic transmission in mossy fiber boutons, a potentially novel mechanism for Munc13s in LTP. Further experimentation is needed to better understand the role of Munc13s in neuropeptide signaling in mossy fiber synapses. Furthermore, I postulate that DCVs likely undergo a basal level of fusion at rest based on the accumulation of DCVs in Munc13-deficient mossy fiber synapses. However, extensive experimentation would be necessary to test whether DCVs fuse in mossy fiber boutons under basal conditions. The speculations could be tested along similar lines as DCV fusion experimentation previously performed in dissociated hippocampal cultures (van de Bospoort et al., 2012; Farina et al., 2015), however be carried out in organotypic slice cultures. By using genetic targeting of granule cells (Kohara et al., 2014), coupled with optogenetic stimulation (Madisen et al., 2012), and phluorin-labeled DCVs (van de Bospoort et al., 2012) one could explore the parameters necessary to evoke DCV fusion in mossy fiber boutons and start examining the molecular machinery regulating synaptic release.

DCVs in mossy fiber synapses have varying morphological characteristics; some DCVs have diameters similar to those of synaptic vesicles, the electron dense material in the center is sometimes segregated to the center of the vesicle with varying halo sizes between the vesicle membrane and the electron dense core, and at times the core is eccentrically located within the DCV lumen. It is possible that the differences in size and dense core opacity are caused by piecemeal degranulation of individual DCVs, as has been described in mouse mast and chromaffin cells, as well as thalamic and hypothalamic neurons (Crivellato et al., 2005). The neuropeptide content of each DCV is not known in mossy fiber boutons. It is also unknown if there is an ultrastructural correlate to neuropeptide content in individual DCVs; if for example, BDNF could be contained in DCVs with a halo, whereas enkephalin or dynorphin are packaged in DCVs with no halo. Both morphologies were observed in mossy fiber boutons. Further work is needed to correlate the morphological characteristics of DCVs in mossy fiber synapses and the neuropeptide content stored within each DCV. If there is a correlation between DCV morphology and neuropeptide content, one could make further comparisons of DCV accumulation at mossy fiber active zones and functional aspects of the synapse.

4.5.4. Presynaptic cAMP and DCV docking

In the present study, I applied forskolin to hippocampal slice cultures to pharmacologically enhance release probability in mossy fiber synapses. Forskolin activates AC1, which in turn increases presynaptic cAMP production and enhances synaptic transmission via downstream effectors (Trudeau et al., 1996; Villacres et al., 1998; Weisskopf et al., 1994). After acute treatment with forskolin, I found an increase in DCV docking, however no change in the spatial density of DCVs within 40 nm of the active zone membrane, indicating a cAMP-dependent redistribution of membrane proximal DCVs. Previous work in *Caenorhabditis elegans* demonstrated that elevating presynaptic cAMP levels with a bacterial light-activated adenylate cyclase increased DCV fusion at the neuromuscular junction (Steuer Costa et al., 2017). Work from the present study supports the idea that DCV docking is influenced by forskolin-induced production of presynaptic cAMP in the mossy fiber synapse but direct evidence of DCV fusion is lacking. Verhage and colleagues showed that DCV exocytosis from dissociated rodent hippocampal neurons requires a large, global increase of calcium ions in the presynaptic bouton that occurs with persistent opening of presynaptic calcium channels during high synaptic activity (Verhage et al., 1991). It is possible that the increase I observed in membrane-docked DCVs is due to downstream effectors of presynaptic cAMP acting independently of action-potential driven changes in presynaptic calcium.

The cAMP signaling pathway has been extensively studied in many cell types including neurons (Seino and Shibasaki, 2005). Two major downstream targets of cAMP in neurons are PKA and Epacs (Evans and Morgan, 2003; Robichaux and Cheng, 2018; Shibasaki et al., 2007). Activation of PKA and Epac depends on the intracellular concentration of cAMP; PKA has a relatively high affinity for cAMP and requires low nanomolar concentrations, whereas Epac binds to cAMP with lower affinity at concentrations at which PKA is saturated (Seino and Shibasaki, 2005). No studies have explored the effects of PKA or Epac on DCV fusion in neurons, however several studies have shown effects of PKA and Epac on secreting cells. Epac, but not PKA, has been implicated in increased secretion of insulin-containing granules from pancreatic beta cells in human tissue (Kang et al., 2003). In mice, PKA-dependent and -independent mechanisms of cAMP-induced insulin secretion from beta cells have been described (Eliasson et al., 2003). In this context, to dissect which pathway, PKA- or Epac-dependent, is involved in the forskolin-induced increase in DCV docking could provide insight into the signaling cascades required for DCV fusion in mossy fiber synapses.

In summary, forskolin-induced presynaptic cAMP in hippocampal mossy fiber boutons changes the membrane-proximal distribution of DCVs; increased cAMP leads to increased DCV docking and DCG-IV-induced reduction in cAMP results in fewer docked DCVs. Presynaptic cAMP is part of the pathway utilized to enhance synaptic transmission in mossy fiber synapses during sustained activity. Neuropeptides, packaged in DCVs, are also implicated in modulating mossy fiber synaptic transmission. This cAMP-induced phenomenon could be a novel mechanism for DCV docking in hippocampal mossy fiber synapses.

4.6. Development of Schaffer collateral and mossy fiber synapses in slice culture

Synapse development is incomplete at birth. Indeed, structural and functional changes occur in animals such as rodents throughout the life time of the animal. In the present study, I investigated the ultrastructure of mossy fiber and Schaffer collateral synapses at two developmental time points (DIV14 and DIV28) in the mouse hippocampus. My data demonstrate a developmental increase in the spatial density of docked synaptic vesicles at both Schaffer collateral and mossy fiber active zones between DIV14 and DIV28.

Mossy fiber and Schaffer collateral synapses undergo structural and functional maturation between the ages of P3 and P21 (Amaral and Dent, 1981; Battistin and Cherubini, 1994; Buchs et al., 1993; Helassa et al., 2018; Ho et al., 2007; Hussain and Carpenter, 2001; LaVail and Wolf, 1973; Marchal and Mulle, 2004; Mori-Kawakami et al., 2003; Münster-Wandowski et al., 2013; Rose et al., 2013; Schmitz et al., 2003; De Simoni et al., 2003; Wilke et al., 2013). Structural maturation of mossy fiber boutons in cultured slices closely resembles the time course described *in vivo* (Amaral and Dent, 1981; Galimberti et al., 2006; Wilke et al., 2013). Since previous studies have concluded that presynaptic structural maturation of mossy fiber boutons peaks at P14, with only minor changes beyond that, including a pruning of filopodial extensions and an increase in bouton volume (Amaral and Dent, 1981; Galimberti et al., 2006; Wilke et al., 2013), my observations yield novel insight into mossy fiber synapse development at the level of fine structural organization at the active zone. Synaptogenesis and spine maturation in CA1 pyramidal neurons has been described (Fiala et al., 1998; Harris and Weinberg, 2012; Harris et al., 1992). Schaffer collateral spines undergo structural maturation from P1 to P40 in both the proportion of spine morphologies and presence of specialized organelles such as spine apparatuses (Fiala et al., 1998; Harris and Weinberg, 2012; Harris et al., 1992; De Simoni et al., 2003). By P12 in rats, most excitatory synapses in the *stratum lucidum* of the CA1 are located at either the dendritic shaft, or the tip of postsynaptic spines (Fiala et al., 1998). Further, from P15 to adult (P70) in the rat hippocampus, spine morphologies change in their relative proportion and density (Harris et al., 1992). Harris and colleagues found a higher proportion of thin spines and, inversely, a lower proportion of stubby spines in the CA1 at P70 compared to P15 in rats (Harris et al., 1992). The heterogeneity of spine morphologies in the CA1 has been postulated to influence structural plasticity during high-activity states, however this remains an open question (Harris and Weinberg, 2012).

The developmental increase in the spatial density of docked synaptic vesicles in Schaffer collateral and mossy fiber synapses was not paralleled by corresponding changes in the efficacy of neurotransmitter release as measured by patch clamp electrophysiology. Neither mean evoked EPSC amplitudes nor paired-pulse ratios changed significantly during this developmental window. This observation was not entirely unexpected, since unlike structural development, the functional changes in both Schaffer collateral and mossy fiber synapses observed in past studies occurs mostly before P14 and after P28 *in vivo* (Battistin and Cherubini, 1994; Ho et al., 2007; Hussain and Carpenter, 2001; Marchal and Mulle, 2004; Mori-Kawakami et al., 2003; Rose et al., 2013; Schiess et al., 2010; Schmitz et al., 2003). In acute rat slices, mossy fiber synapses exhibit similar paired-pulse facilitation and post-tetanic potentiation between P10 and P20 (Hussain and Carpenter, 2001). However, between P14 and P21 mossy fiber synapses in acute mouse slices exhibit increased frequency facilitation that is not dependent on postsynaptic receptors (Marchal and Mulle, 2004). Developmental changes in Schaffer collateral synaptic function in acute slices also occur before P14 and after P28 in rodents (Hussain and Carpenter, 2001; Schiess et al., 2010).

Understanding why the observed increase in the spatial density of docked synaptic vesicles does not translate to detectable changes in postsynaptic responses requires the consideration of the complex interplay between multiple factors. These factors include presynaptic calcium channels and buffers, as well as developmental structural and molecular changes in the postsynaptic compartment, and in the abundance and properties of postsynaptic transmitter receptors.

Modifications to presynaptic calcium channels such as the relative proportion of VGCC subtype can shape presynaptic function. Spontaneous and evoked synaptic currents increase during development in dissociated hippocampal neurons (Basarsky et al., 1994). Dissociated cultured neurons had a developmental increase in calcium influx that coincided with an increase of the proportion of P/Q calcium channel expression as the neurons matured (Basarsky et al., 1994). At mossy fiber synapses (at P20-23), the number of P/Q-type calcium channels estimated per active zone is approximately 23 (Vyleta and Jonas, 2014). It is, however, not known how this number is developmentally regulated.

Developmental changes in the spatial distribution of presynaptic calcium channels have been described in the rat cerebellum (Miki et al., 2017). Between P14 and P28, calcium channels decrease in number and form organized clusters that preferentially aggregate at the periphery of the active zone (Miki et al., 2017). Consequently, Miki and colleagues found that the number of docked vesicles correlated with the number of calcium channel clusters indicating fewer vesicles were docked at active zones at P28 (Miki et al., 2017).

Bornschein and colleagues found that there was a developmental decrease in the calcium channel-sensor coupling distance in somatosensory layer-V pyramidal neurons from wild-type mice between the ages of P10 and P24 (Bornschein et al., 2019). Despite tighter coupling, there was no change in the RRP and release probability between the two ages (Bornschein et al., 2019), indicating that developmental changes can occur without affecting the functional output of a given synapse. In mossy fiber synapses, calcium channel-synaptic vesicle coupling distance is around 70 nm, much larger than other synapses with a higher release probability (~20 nm in the Calyx of Held) (Chen et al., 2015). An unanswered question is whether mossy fiber or Schaffer collateral synapses have a developmental change in calcium channel-sensor coupling distance.

The unchanged functional output of both mossy fiber and Schaffer collateral synapses could also be due to an increase in endogenous calcium buffers (Blatow et al., 2003; Luiten et al., 1994). For example, calbindin is a fast-acting endogenous calcium buffer that influences synaptic functional properties in mossy fiber synapses (Blatow et al., 2003; Dumas et al., 2004). In the developing rat hippocampus, calbindin expression steadily increases in granule cells and mossy fiber synapses from P5 to P20 (Luiten et al., 1994).

Presynaptic mitochondria can rapidly sequester presynaptic calcium in synapses to maintain a low release probability (Kwon et al., 2016). Presynaptic mitochondria are more abundant in mossy fiber boutons compared to Schaffer collateral synapses (Amaral and Dent, 1981; Shepherd and Harris, 1998; Smith et al., 2016). Indeed, as assessed by 2D transmission EM, in the untreated slice cultures at DIV28, nearly every mossy fiber bouton contained at least one mitochondrion whereas approximately one third of Schaffer collateral synapses harbored a presynaptic mitochondrion (see Figure 34 in the appendix). However, developmental changes in the presence of presynaptic mitochondria are poorly understood.

Postsynaptic development includes changes in glutamate receptor subtype expression, and in the localization of receptors in relation to synaptic release sites (Ho et al., 2007; Marchal and Mulle, 2004; Sans et al., 1996). For example, CA3 pyramidal neurons undergo a developmental increase in kanaite receptor expression at the mossy fiber-CA3 postsynaptic compartment that marks the onset of frequency facilitation and potentiation observed at these synapses (Marchal and Mulle, 2004). In another study, expression of calcium-permeable AMPA receptors underlies early post-natal long-term depression exhibited at mossy fiber-CA3 synapses (Ho et al., 2007). In Schaffer collateral synapses, Sans and colleagues found an increase in the expression of proteins associated with the postsynaptic density involved in anchoring NMDA receptors at the synaptic junction (Sans et al., 1996). They postulated that during development, the accumulation of NMDA-anchoring molecules helps localize NMDA receptors closer to release sites (Sans et al., 1996).

The factors outlined above, perhaps in combination, may occlude the detection of functional changes associated with my observed developmental changes in vesicle docking. As described in other synapses, the structural alteration in presynaptic vesicle organization may be silent, exerting no influence on release probability or on short-term facilitation (Bornschein et al., 2019).

5. Conclusion

In the present study, I performed a comparative ultrastructural analysis of the functionally distinct hippocampal Schaffer collateral and mossy fiber synapses to assess whether the availability of docked and primed synaptic vesicles contributes to differences in release probability. To address this question, I combined hippocampal slice culture, HPF, AFS, and electron tomography to accurately resolve the organization of vesicles at presynaptic active zones.

My ultrastructural analyses revealed mossy fiber synapses, which functionally exhibit a low release probability and robust synaptic facilitation, harbored fewer morphologically docked synaptic vesicles than Schaffer collateral synapses. This supports the hypothesis that differences in synaptic release probability depend on the availability of fusion-competent synaptic vesicles. Additionally, mossy fiber synapses had a prominent second, membrane proximal pool of vesicles that represent a LS state of tethered synaptic vesicles, which are ideally positioned to rapidly replenish the RRP during periods of increased synaptic activity. This membrane-proximal accumulation was not observed in Schaffer collateral synapses.

Additionally, my work identified, for the first time, that three morphologically distinct types of vesicles dock at mossy fiber active zones: synaptic vesicles, giant vesicles (clear-core vesicles with a diameter exceeding 60 nm), and DCVs. All three vesicle types docked at mossy fiber synapses in a Munc13-dependent manner. Past studies reported the presence of giant vesicles and postulated they were neurotransmitter-filled vesicles responsible for giant monoquantal events observed in CA3 pyramidal neurons (Henze et al., 2002b; Laatsch and Cowan, 1966; Rollenhagen et al., 2007). I found that giant vesicles docked exclusively at mossy fiber active zones in the same proportion as giant mEPSC events of mossy fiber origin measured in slice culture. Giant vesicles were not merely by-products of endocytosis because they persisted in acute, pharmacological silencing of network activity and in constitutively silent, Munc13-deficient mossy fiber synapses. Evidence suggests that giant vesicles may be trafficked from granule cell somas, however many questions about giant vesicle identity and origin remain.

I performed a quantitative analysis of respective vesicle pools at mossy fiber active zones and compared my data with published functional estimates of the RRP. I demonstrated a considerable overlap between the total numbers of morphologically docked and functionally primed and fusion-competent synaptic vesicles and proposed that larger reported RRP estimates involved the recruitment of membrane-proximal vesicles.

I examined whether changes in synaptic release probability induced by acute pharmacological manipulations of presynaptic cAMP trigger corresponding changes in vesicle organization. My data reveal that DCV docking is particularly sensitive to changes in presynaptic cAMP levels. Surprisingly, I observed no change in the organization of docked synaptic or giant vesicles after increasing presynaptic cAMP. These findings support a view in which mechanisms mediating cAMP-dependent potentiation of mossy fiber synaptic transmission operate downstream of synaptic and giant vesicle docking. Additionally, they highlight the potential modulatory role of DCV-mediated neuropeptide release in mossy fiber plasticity processes.

In conclusion, my work further supports the hypothesis that initial release probability is determined by the availability of docked and primed vesicles and that the structural organization of vesicles at active zone release sites can provide insight into presynaptic functional properties. Furthermore, it emphasizes that methodical and thorough high-resolution ultrastructural analyses are useful to reveal novel insight into ultrastructure-function relationships in other synapse types in the brain.